Simulating the Mind II

Dietmar Dietrich • Volker Hartmann Cardelle

Simulating the Mind II

From Artificial Intelligence to Neurology
and Psychoanalysis

 Springer

Dietmar Dietrich
TU Wien
Vienna, Austria

Volker Hartmann Cardelle
Ghent University
Ghent, Belgium

ISBN 978-3-031-69529-2   ISBN 978-3-031-69530-8   (eBook)
https://doi.org/10.1007/978-3-031-69530-8

Translation from the German language edition: "Simulating the Mind II - Psychoanalyse, Neurologie, Künstliche Intelligenz: ein Modell" by Dietmar Dietrich, © Shaker Verlag 2021. Published by Shaker Verlag. All Rights Reserved.

This Springer imprint is published by the registered company Springer Nature Switzerland AG
The registered company address is: Gewerbestrasse 11, 6330 Cham, Switzerland

If disposing of this product, please recycle the paper.

With the support of
Klaus Doblhammer and Dorothee Dietrich

Foreword of Two Engineering Scientists

Vienna, 20 years ago, a meeting room at TU Vienna's Institute of Computer Technology: a group of psychoanalysts, neurologists, computer scientists, and computer engineers (note the difference!) discuss an impossible project: to replicate (some aspects of) the human mind/psyche in a machine. This meeting was a clash of cultures, if not entire worldviews, but everyone also felt the great potential and the enormous amount of knowledge that eventually could be gained from the respective "other side" of the group. Together we created an inspiring scientific ecosystem that allowed for radical ideas.

What followed were years of fundamental research with a roller coaster of failure and success. The driving force that kept everyone motivated was the idea that it was indeed possible to bridge the gap between the material and the informational and to create a model that describes how to equip inanimate objects with mental capabilities, making technical devices ultimately compatible with their creators. We foresaw applications in transport, automation, energy, robotics, and medicine where machinery—in its broadest sense—would be on a par with humans, forming a cooperative whole. Twenty years later we are still not there, but the motivation still holds.

The last handful of years have been dominated by the success and impact of deep learning: Large language models and data science have invaded and disrupted virtually all corners of our society. However, once machines and people are involved, AI should be explainable, fair, and trustworthy. We do not want a nuclear power station, or an autonomous vehicle operated by a mysterious black-box model as is the case with deep neural networks. Engineers therefore combine machine learning with heuristic problem knowledge and structure, leading to "physics-informed" AI. Differential equations and topologies of power grids are for instance expressed as graph neural networks to model complex power systems.

This refreshing, educating, and inspiring book takes this very route: Structure and functions, inspired by psychoanalysis, help to create smarter artifacts. And let's be honest: You can pour as many—or even more—artificial neurons as human brains have biological ones into a bucket, but no mind will emerge. It will not say "Mama!" and start questioning the meaning of life. A hierarchical, functional structure is needed.

Delft University of Technology, Delft,   Brigitte Palensky
Netherlands   Peter Palensky
July 2024

Foreword of a Medical and Psychoanalytical Scientist

The examination of complex processes, affect-cognitive functioning, behavioral observations, and the significance of unconscious interpersonal processes for mental development and mental health are highly relevant—not only in a clinical context, but also for the understanding of the brain in neuroscientific research.

In recent years, neuroscience has "increasingly endeavored to capture the processes described in psychoanalytic findings in terms such as containment and reverie from psychoanalysis. Psychoanalysis offers a mature body of theory and knowledge for understanding intrapersonal and interpersonal processes. Many concepts include clinical observations and are operationalized for scientific examination. However, scientific progress is always related to the quality of the research instrument. This book offers an opportunity to link different levels of observation, theory, and scientific knowledge processes. This transdisciplinary work is extremely relevant and essential for research into individualized treatment processes in order to do justice to the complexity of human beings. Let me illustrate this with an example.

Pain-avoiding behavior can be observed, the subjective meaning of the pain can be discussed and elaborated with the patient, and coping with the pain can be achieved through clarification, confrontation, interpretation, and working through. The concepts of containment and reverie describe part of the analyst's activity. In neuroscientific studies, the concept of the default mode network (DMN) is increasingly coming into focus. In scientific terms, this network describes a specific brain function that enables the brain to learn, think, and plan. The DMN comprises those brain regions that show high activity during resting phases, while it decreases during goal-oriented cognition. This suggests that the DMN is associated with self-related mental processes. In line with psychoanalytic concepts of the ego/secondary process and the id/primary process, observations indicate that the default mode network is significantly involved in the regulation of entropy. This regulation aims at the thinking of the primary processes, and as a result of this regulation free energy/ entropy is minimized. Empirical data show that painful, traumatic experiences, for example, lead to a long-lasting reduction in functional connectivity within the DMN, which corresponds with the increased use of primitive defense mechanisms in connection with painful, traumatic experiences. Defense mechanisms such as projective identification, dissociation, and fragmentation may aim to reduce or bind excessive amounts of free energy resulting from the traumatic experience. However, these rather archaic attempts to minimize error signals are often ineffective and lead to compulsive patterns of behavior that unconsciously perpetuate the pain-causing traumatic past environment.

This complexity poses a challenge for therapeutic practice and research into such processes and requires further investigation into concepts such as reverie and containment, including symbolic information theory. Advances in this area of research could provide valuable insights into the specific processes of therapy and thus the basis for effective psychotherapeutic practice. The parallelization of scientifically describable phenomena (language of neuropsychiatry, psychoanalysis, physiology, medicine, physics) with the language of computer science (language of logic) is an optimal opportunity for a cooperative way of working to make behavior, meaning, function, and structure further researchable. The book shows a future-oriented path of development.

Medical University Vienna, Vienna,   Henriette Löffler-Stastka
Austria
July 2024

Prologue

Since my youth, one subject of inquiry has caught my interest more than any other: How does our brain work? Why am I aware of myself? How is the relation between my consciousness and my feeling to be understood? What is a feeling? For thousands of years, humans made a strict distinction between their spirit and soul on the one hand and the body on the other. The body was only the carrier of the spirit and the soul, which left the body when the person died. But where were the spirit and soul before birth? Slowly, humans began to understand that spirit and body belong together and cannot be separated from each other and that the immortal soul is a figment of man's imagination that cannot be related to natural science. It is the subject of religious teachings.

But how is the connection between body and spirit to be seen?

These are questions that I could hardly ask at school or at university without being looked at suspiciously. You quickly offend religious people with such questions, or you are classified as a nerd, as a stubborn materialist, or, in my student days, even as a communist. As a consequence—although I had already visited a few seminars in philosophy—I suppressed my urge to find an answer to these questions, i.e., to understand the mental part and its connection to its body, and instead studied electrical engineering, a subject that would quickly allow me to provide for a family. For my diploma thesis, however, I allowed myself another attempt to approach this question. I chose the topic of how neurons can be replicated electrically and even simulated with a computer, based on the idea of how information processing *functions* in a network of neurons. How do neurons "calculate"? In them, information is represented by the timely distances of electrical impulses, regardless of the source of the information, be it the eye, the nose, the hands, or the back. The brain only receives these electrical impulses and their different timely distances and out of these it forms feelings and consciousness. How is this possible? How does this happen?

Even before the year 2000, as a professor of computer technology, I found young students and assistants who were willing to conduct with me research in this field and to look for completely new approaches that were different than those that had been pursued so far—although they evidently could not lead to the desired goal. Thanks to research projects in areas such as air traffic control, energy technology, building automation, and chip design, I more and more was able to raise the necessary funds so that we became less dependent on research grant applications. For the most part, research proposals did not bear much fruit. Research projects, as I had them in mind and as I spelled them out, were always sent to reviewers from engineering, neurology, psychiatry, or psychology. And these reviewers had to reject our proposals, since they expected mathematical algorithms or behavioral models or classical artificial intelligence approaches. They could not—as far as I can tell from the reviewer's comments—relate to my ideas of *functional models of psychoanalysis and neurology* that were built upon an extended computational theory of Georg H. Mealy. Nevertheless, the numerous dissertations—which usually took more than 5 years to be completed—step by step brought about results that indicated that we were and are on the right track. The problem, however, is that you have to seriously deal with *psychoanalysis*, which is difficult for technicians, and at the same time you need to have a solid understanding of *computer technology*. Here, I deliberately speak of computer technology and not computer science, because our approach is not about algorithms, which are the main content of computer science. It is primarily about understanding the connection between the description of a computer by means of *physical methods* and the description by means of *information methods*. For me, this is the pivotal theme of computer technology as I have always understood it. Especially if one is intensively involved with the design of microprocessors—as I was—and knows what micro-programs are and how they can be described[1] and implemented on the one hand in terms of hardware—i.e., approaching from matter— and on the other hand in terms of software—i.e., approaching from engineering. Understanding these two large fields takes time. And the necessity to describe a system with the methods and the laws of physics, chemistry, and neurology, and at the same to work with the methods and the laws of engineering/information theory[2] while relying on the models of psychoanalysis, is extremely demanding and sets a high hurdle for everyone who wishes to understand this approach. After all, these scientific fields are completely different worlds that nevertheless have to be joined together. But once you get fully into it, it is great fun—and even more: You realize that you have gained a better understanding of the world.

However, everyone must be aware that the subject is demanding. You cannot get into this approach with the attitude for which in German there is the phrase:

---

[1] The term *describe* is to be understood in the scientific sense, i.e., a description with the aid of scientific methods such as mathematics, the axiomatic use of terms, etc.

[2] In engineering sciences and technology, terms such as *engineering* and *information theory* are generally applied narrowly to contemporary technically realized systems. In order to make new concepts possible, we have to overcome such narrow conceptualizations. Therefore, in the following I will use the term *information theory* not only for the classical theory, but also for all other theories concerning the transport, processing, and storage of information. I hope that this conceptual expansion will become clear in this book.

> Bathe me, but don't get me wet!

In relation to this book, the attitude for artificial intelligence engineers could be spelled out as:

> Explain to me the coupling between the nervous system and the psyche, but don't bother me with the science of psychoanalysis.

For psychoanalysts, psychologists, and pedagogues the sentence can be translated into:

> Explain to me the coupling between the nervous system and the psyche, but do not bother me with electricity and voltage and especially not with computer technology.

One needs to overcome some intellectual resistances, if one wishes to get into this approach. However, the good news is that the content of the book does not presuppose a profound knowledge of natural science or psychoanalysis, but attempts to introduce interested people of all disciplines to the subject of how to view *body* and the *mental part* as a unity on a natural scientific basis and with a general scientific interest and understanding. Complex topics cannot be explained meaningfully and convincingly without at least a minimum of basic understanding of the sciences involved. This basic understanding must be acquired at the beginning in order to be able to dive deeper into the still relatively new topic of the bridge between the neurological and the psychic description.

Preface by Dietmar Dietrich

All thinking is based on this wrinkled, gray structure we call the brain. All decisions are made with it. It acts, reacts, and performs complex actions. And yet, to this day, we can scientifically only comprehend, model, and understand it to a very limited extent.

So, we make decisions with a "thing" that we do not really understand? When you think about it, this is baffling. I am *me* through this thing. Through this thing you are *you*. This "thing" should be our most important goal of research. Interestingly enough, it is exactly not. A lot of time and money is invested in genetic engineering and neurology—of course not as much as in the gaming or pharmacological industries—but research into the mental apparatus in the context of the neurological part, i.e., modeling the brain in a holistic way, is sparse in every respect. Jeff Hawkins already noted this (Hawkins 2004, pp. 7–8), but not much more has happened since then; I would say that a scientific breakthrough has not happened in classical artificial intelligence.[3] The results of the scientific project SiMA (Dietrich et al. 2015) show that the research of this organ[4]—and in our

---

[3] Since the publication of the German edition of this book, ChatGPT has been developed, which certainly represents a quantum leap in AI. However, one has to keep in mind that this has been achieved mainly by an additional learning function and corresponding algorithms. This is a different approach from the one I am taking. Its enormous limitation lies in the fact that it does not take into account and even less so emulate the structure of information-functions and of hierarchical information valuation that form the basis of our psychic apparatus. This is, of course, a very difficult task to tackle first from the point of view of basic research, which is why many do not even approach it. The goal of this book is to explain how this can be achieved.

[4] In the beginning of this book, for the sake of simplicity, the expression "brain" is often used instead of "nervous system" or the synonymous term "neuronal system." Crucially, for the purposes of the book, terms like those subsume both the physiological part (the neurons) and the psychic apparatus. However, in the SiMA project (Dietrich et al. 2015), on which this book is based, it was recognized very early on that differentiation of such expressions is important, so they are discussed and defined in detail throughout the book. It is in this sense that we must understand the terms nervous system and neuronal system in this book. In the medical field, the term nervous system

understanding this includes the mental apparatus—as a whole is quite feasible today. This is the message of this book. In it many topics are addressed, topics such as artificial intelligence, brain, psychoanalysis, behavioral model, complexity, bionics of the mental, consciousness and awareness, preconscious and unconscious, the functional—not only behavioral—description and modeling of the brain and especially of the mental apparatus, and some more.

The book is deliberately written in a rather general but also scientific way. I want to convey a vision, and I deliberately write in a personal manner to be able to convey the idea that this technology must be based on psychoanalysis. The goal is to reach as many readers as possible, to enthuse them for the topic, to discuss it, and also to provoke. We have to break with traditions in order to give space to new forms of thinking and to understand them. It is crucial to take a new path in automation, in artificial intelligence, and in (technical) cognitive science, but also in psychoanalysis and neurology, in order to create a common basis for scientific and technical possibilities that have been previously inaccessible to us.

I discussed a lot about the title of the book. We in the SiMA team have been shaped by experiences of rejection on conferences. In the world of technology, there is little interest in psychoanalysis. In the world of psychoanalysis, people usually refuse to deal with mathematics and technology, especially computer technology. This is how I came up with the provocative title (which in the end seemed too narrow to me): "The fourth blow of mankind." Sigmund Freud pointed out three basic narcissistic blows mankind had suffered so far: the cosmological, the Darwinian, and the psychological blow. Is there another blow that comes from information theory or computer technology? Or is there an information theory of the brain on the one hand and an information theory of computer technology on the other hand? Let us try once on both sides, on the side of engineering science and the science of psychoanalysis, to forget for a moment the prejudices and to respect the knowledge of the other side. One will perhaps understand that natural science must in principle use the same physics, chemistry, and biology and also the same information theory, also for the understanding of the brain. One will also have to recognize that physics and chemistry obey different laws and methods than those of information theory. And one will have to recognize that the brain can only be modeled and thus simulated comprehensively and holistically by natural science, if one grants the appropriate space to both fields of science.

At the beginning of the book project, I discussed with those who gave me crucial support and who are also named on the cover page whether we should write in the "we" form or I should write in the "I" form. We decided on the first-person form. There are several reasons for this. The initiator of the project and the book is me, one of the authors: Dietmar Dietrich. But, as explained in detail in the book, I could not write the book alone because of the scientific methodology chosen and sought fellow

---

(or similar terms) is mainly used, and in the technical field, the term neuronal system is used. For this reason, the book tries to use the term nervous system when it deals more with medical aspects, and the term neuronal is preferred when the topic is more related to modeling or technical aspects.

contributors to cover the psychoanalytic and philosophical aspects of the subject. The mindset underlying the project, which is necessary to be able to model the brain, should be consistently applied in the book as well. And that makes it mandatory to involve experts from the various fields. Nevertheless, this book should not be a collection of different essays [like the first book of the SiMA project (Dietrich et al. 2009)], but the text should be a single cast. This means, however, that the team must agree on one opinion in all points, because this is the only way to obtain a conceptual world free of contradictions (called axiomatics in natural science), which will be treated in detail.

Another point must be addressed: I not only initiated the project, but also helped shape it over more than 20 years. Personal epiphanies have not infrequently pointed me in the correct direction. These emotional moments should also be conveyed. Perhaps this will enable the reader to understand certain aspects more easily. Thus, I purposefully do not want to write soberly and objectively; we want to evoke emotions. Maybe this incites some readers to write us their opinion and their questions, which might lead to a broader discourse.

And please, do not recoil from the second chapter. Yes, I want to be provocative in that one. I already mentioned it. But I promise I will answer all the questions and hypotheses raised in it on a natural scientific basis. I will try not to leave any item open unless I explicitly point out that I do not have an answer to it.

Berlin, Germany   Dietmar Dietrich

Preface by Volker Hartmann Cardelle

The reader might be surprised to find that the translator of this book is not a native speaker of English, as it is commonly the case. I also encountered some skeptical reactions from people close to me when I told them that I had decided to assail the task of translating this book. There are two major reasons why I dared to take up this challenge but also believe that this way of translation actually has clear advantages over the traditional way. The first is of technical nature: The advent of DeepL and its considerable increase in quality and usability in the last few years, which improved even after I started working on the translation and I must admit that those improvements were vital. Many people, so I believe, still underestimate the potential of this technology. It has reduced the difficulty of finding a proper formulation in the target language massively to say the least. It is this part of the work where a native speaker trumps a foreigner, I believe. And DeepL reduces this advantage of effort. Still, I think that DeepL cannot do away with the problem that a foreigner lacks the sense for the expressions, especially if he does not live in a country were the target language is spoken, as it happens in my case. Hence, I believe that despite the best effort the following text is not free of Germanisms that are a stain on the verbal style of this book. This then brings me to the second reason that has more to do with the theory of translation and my understanding of translation. I firmly believe that if a translation contains traces of the original language—Germanisms in this case—this can actually be an advantage (as long as they do not happen too often, that is), because they display the fact that this book is not the original text but a translation—a fact that tends to be hidden behind the veil of an all too good and too familiar style of verbal expression. And this veil then easily lures the reader to take the text at face value, with which I mean that it is forgotten that a translation is not a copy of a text in another language (as if languages were just different codes) but is rather an interpretation of the text in another language. And like any interpretation it can be spot on and highly illuminating, as well as totally lost and confusing, and it might even be distorting the meaning of the original text. Therefore, in my conviction the most important factor for judging the quality of a translation is its ability of unveiling the thoughts that are carried in the original text and not how familiar the verbal style of expression sounds to the reader. Therefore, it is more important for the translator to properly understand the thinking behind the original text than being a native speaker of the target language. This then brings me to the content of this book: Dietmar Dietrich's thinking, i.e., the thinking behind the project SiMA, and therefore to how I had the pleasure to get to know him.

When I first met him in Chicago in 2016 at the conference of the Neuropsychoanalytic Society (NPSA) I was told that he was somehow interested in using Sigmund Freud's theories for the simulation of human psychic apparatus. I was highly intrigued about that but, and this I must not conceal, I also had the arrogant belief that I, having studied intensively Sigmund Freud's entire works over 6 years before, might be in the position to explain to him something about Sigmund Freud—I was mistaken. In the intense conversation that quickly evolved at one point I was totally stunned, when I suddenly became aware that Dietmar had a brilliant solution for the problem I had been wrestling with intensively for at least 3 years: how to understand Freud's conception of the "psychic apparatus" (Freud 2017, p. 406) and how it could be related to the brain. It was not only a solution; it was *the* solution. I was and became more and more sure about this, since suddenly, by relying on Dietmar's explanations, all the passages of Sigmund Freud that had puzzled me were mentally orchestrated into one consistent display of thought. I became aware that until then my understanding of the psychic apparatus had been too physiologically oriented—a judgment about my thinking that I should have to repeat as a result of conversations with Dietmar quite a few times in the years to come. This was comically mind-blowing to me, since I was the student of psychoanalysis and he the professor of engineering: I was supposed to be the advocate of the sovereignty of the psychic apparatus, yet he championed it in a manner that was awe-inspiring. I remember that, as a consequence of our encounter, I boldly declared at the post-conference meeting that I thought to have come to understand why Sigmund Freud had abandoned his development of a model of the functioning of the brain to account for mental phenomena as he had started to develop in the *Entwurf einer Psychologie (1895)* (Freud 1999). And although I was convinced that I had found somebody who actually had the solution to the riddles of metapsychology and provided a way of connecting metapsychology with the rest of natural science—something I had dreamed about since I had started studying Freud. Thus, I was highly motivated to learn from him, and yet it took me almost five long years with numerous exchanges of long emails, many video calls, two further personal meetings, and my work as corrector of the original version of this book until I reached the point that I was able to say with conviction that I had understood the fundamental theory on which SiMA relies. Until that point it happened over and over again that I misunderstood certain aspects of the theory and model. And it was not before 2023 that I was able to provide written elaboration for the bold claim I had made in 2016 in Chicago in the form of a scientific contribution written together with Dietmar Dietrich (Hartmann Cardelle and Dietrich 2022).

This long intellectual travel made me arrive at the conviction that if anyone wants to provide a good translation, they first and foremost must properly understand Dietmar Dietrich's thinking. Without this understanding the translator would repeatedly stumble over misunderstandings, as it happened to me time after time in the course of my learning years with Dietmar. And of what use is a translation if it is written in an elegant language but it cannot convey to you the thinking of the author? And since this thinking is so hard to get into and especially its application to psychoanalysis is understood by so few, I might actually be the best translator out there for this book—a book that also in the original does not stand out for the elegance of its prose, but which transports a thrilling and illuminating thinking. I must nevertheless prepare the reader that they will need some patience to work themselves into this book since basically all positions that are brought together by Dietmar will find some statement that most likely will appear provoking at first glance—thus one needs the patience of taking a second look at different thoughts.

Just before I conclude I want to address the typical commonplace between psychoanalysts that might be looming: Yes, this thinking has been to this day many times highly valuable for my clinical work.

To conclude, I can only invite the readers to make the effort of working themselves into this book and hope that the invitation is followed.

Chur, Switzerland   Volker Hartmann Cardelle

Acknowledgments

Without question the knowledge on which this book is based has emerged from the many controversial discussions of the SiMA project. SiMA stands for "Simulation of the Mental Apparatus and Applications"; earlier project names were Intelligente Küche (Intelligent Kitchen) and ARS (Artificial Recognition System). Since I initiated the project in 1999 I have employed numerous people from different disciplines in it, first and foremost computer engineers and primarily electrical engineers, because they are the right persons to do this. The institute where we worked and which I was largely in charge of at that time was located in the Faculty of Electrical Engineering. In addition, scientists of computer technology, computer scientists, psychoanalysts, psychologists, neuroscientists, psychiatrists, and philosophers were also employed in the project. But most of the work was done by the many PhD students, diploma students, and other students who participated in seminars or wrote their bachelor's theses. In the weekly project meetings, there were often heated and emotional discussions, different methods were tested, controversial opinions were held, but I tried to never lose sight of the goal of *placing psychoanalytic theory on a natural scientific foundation*. This was in order to achieve two things above all else: First, I am convinced that technology has managed to produce excellent methods and principles that should be applied in all natural sciences, including psychoanalysis. Accordingly, the model of the psychic apparatus must be easier to work out and understand with these tools of engineering. Secondly, I always wanted to create a basis for *artificial intelligence*, so that computers and thereby also robots are given a structure of thinking like humans, so that they can take over tasks for us that help us live more comfortably and better.

SiMA is a team effort. The field is simply far too broad and encompasses too many specializations for one person. All of them also had to witness again and again how the project was and still is fiercely attacked from various sides. This is addressed several times in the book. Hence it takes a lot of idealism, and you have to have faith in your knowledge and your own insights to keep pursuing this new path despite major hindrances. And of course, doubts always arise. Resolving doubts takes time, but such a difficult topic cannot be mastered without reflection.

Therefore, once again, I would like to thank everyone—also because I know that I have demanded a lot from all the contributors—for their contribution, which they have made with passion. This also helped me to write this book.

I could never have done it without all the contributors; thus, I am grateful to all those who have supported me in this idea and contributed with commitment. This includes in particular all the mentioned PhD, diploma, and bachelor's students of the SiMA team, but also people who directly or indirectly had a great influence on the project such as (listed alphabetically by letter of Surname): Elisabeth Brainin (psychiatrist, psychoanalyst), Dietmar Bruckner (Dr. techn.), Andreas Cieslik-Eichert (teacher), Martin Fittner (Dr. techn.), Georg Fodor (Dr. med., psychoanalyst, psychiatrist and neurologist), Tobias Gawron-Deutsch (Dr. techn.), Volker Hartmann Cardelle (M.Sc. psychoanalyst), Matthias Jakubec (Dipl.- Ing.) (†), Roland Lang (Dr. techn.), Clemens Muchitsch (Dr. techn.), Brigitte Palensky (Dr. techn.), Peter Palensky (Prof. Dr. techn.), Charlotte Rösener (Dr.-techn.), Thilo Sauter (Prof. Dr. techn.), Roman Widholm (Psychoanalyst), Heimo Zeilinger (Dr.-techn.), and Gerhard Zucker (Dr. techn.).

Contents

Chapter 1
Motivation

In 1956, Marvin Minsky, one of the founding fathers of artificial intelligence (AI), postulated (Rubner 2002) that it was (freely quoting) "... only a matter of time ... before there were robots that could rival humans ...". Does it really look like that, if we consider all the tasks that humans are able to fulfill? In the same way, similar formulations have not become true, such as (also freely quoting) (Simon 1965): "... in 20 years, machines will be able to do everything that humans can do ..." or (allegedly according to Paul Levi again freely quoting) (author unknown 2003): "... in ten years, robots will perfectly comprehend the scenery around them ...". It remains true, however, that automation, with the support of computer technology and computer science, is increasingly not only replacing human labor, but also performing work that requires intelligence to an ever increasing level.[1] This process of growing intelligence in machines is to be welcomed.

I am 78 years old and a seasoned man. It is foreseeable that the time is not far off when I will need help for my daily routine. In Taiwan, where the problem of old age is taken very seriously, a decree has been formulated that all technical universities must deal with the question of how to develop robots "that assist the elderly". And it is fair to say that research institutions around the world are working on this topic. But what research goals have been set? So far, they are simple tasks: to recognize and help when someone has fallen over, or to provide support for taking medication on time. This cannot and must not be all that we aim for. Our research content is aimed in a completely different direction.

Of what use is it if my smart phone rings to remember me to take my medication, but I am distracted and do not want to be disturbed, and thus I simply mute the ringing? How many (particularly older) people behave in exactly this way? We need

---

[1] At this point I have to refrain from defining "intelligence". The explanation and justification of the method is not easy and requires some knowledge of information theory, which can only be developed in later chapters.

D. Dietrich, V. Hartmann Cardelle, *Simulating the Mind II*,
https://doi.org/10.1007/978-3-031-69530-8_1

machines that are so intelligent that they can respond to people individually and not act like programmed automatons.

I call my robot, as I imagine it, Karl. Karl must understand the situation I am in. Do I want to be interrupted? Can I be interrupted? Do I need to be interrupted because it's getting urgent to take my pills? Or can it wait another 10 min or a bit longer? I imagine it like this for my old days: I come home from a party in the evening. Karl greets me, takes my bag from me, helps me out of my coat and first offers me something to drink. Perhaps he also hands me my medication. Then I ask him what has happened in my absence. Did anyone pass by? Were there any urgent calls? Has Karl cooked? Is there enough for my wife, who will be coming soon after me? Of course, Karl would tell me immediately if the faucet was dripping, something got broken or is not working properly. Of course, he would observe whether I had really taken the medication, and if not, carefully point it out to me again more urgently a short time later, until he finally got me to swallow it. But always friendly and not too intrusive. He shows understanding for me as an older person who is no longer so nimble in handling several processes at once. Karl is attuned to my wife and me, a nice household slave who sees his fulfillment in accomplishing his tasks with success, outwitting me now and then, and being praised for his work by us. This requires the robot to have a certain *feeling* for situations. That requires the robot to have sensitivity and, above all, *consciousness*.

To go straight to an embarrassing topic that hardly anyone wants to talk about, but which can affect us all: My friend visited, as he does almost every day, his aged and sometimes confused mother and found her in the apartment on her way between the toilet and the living room. She had forgotten to pull up her underpants. She did not notice them. She had also forgotten to clean herself. In hospitals or old nursing homes, this is now generally the job of young nurses, which is enormously embarrassing, especially for old people, when they have lucid moments in between. The young people get used to it, the old people less so. They have to repress it. I also know this from my mother, who was in need of care. For her, these moments were horribly embarrassing. In her old age, she often cried because of this bitter experience. I sometimes imagine that this would happen to me. One of the young nurses, who in these situations often do not use the correct "you" and instead use the "we", points out to me: "We have to clean our bottom now and pull up our pants." Imagine to be in this situation and comprehend it: Can there be anything more embarrassing? It would be an enormous humiliation for me that I would hardly be able to digest. A robot would perhaps make me realize that I am no longer the master of my senses, but it would certainly not put me to such shame—because it is, after all, a machine.

Or another, completely different example, which I have often presented at conferences: A young, dear neighbor in Brunn am Gebirge, where I lived with my family, had to leave the kitchen for a moment because the front doorbell rang. In the kitchen was her little daughter, who driven by passionate curiosity crawled from the chair to the table, in order to pull the freshly boiled tea to her. She scalded herself so dramatically that she had to be operated six times over many years and the scars can still be seen today. A monitoring system for such mundane situations should be an easy task for an experienced electronics engineer—I thought 24 years ago!

As a tenured professor for computer technology and automation, this did at first not present itself as a real problem for me. In my institute there were many young, highly motivated, intelligent students who, together with my assistants, constantly conjured up the most astonishing inventions with the simplest of means and simple algorithms in their laboratory exercises or bachelor's and diploma theses. Starting with a disk that has to be balanced in such a way that a free-moving ball is kept as good as possible in the center, or an automaton that adjusts the partial surfaces of a Rubik's Cube in minimal time so that all surfaces on a side have a uniform color. I also think of the programming of drones that perform fantastic flight maneuvers. I also had spent many years working in development department of industrial companies. So, what could possibly go wrong?

The humbling realization: to this day we have not yet succeeded in developing a simple monitoring program to detect simple situations in which young children are in danger. And we are not alone in this. Some creative companies have also tried without success, as we learned at conferences. Why is that?

It is known that railroad trains, subways and commuter trains have been running fully automatically under certain conditions for a long time. And there has been talk for a very long time about building self-driving cars, but under what framework of conditions does that actually work? As long as there is a regulated process and the route is readily comprehensible, the various possible behaviors of the vehicles can be planned for and taken into account. A subway always travels the same route. Unforeseen situations are kept within limits, because the rails determine the route. For motor vehicles, the situation is more complicated, but the difficulties should be well under control in the near future. For the motor vehicle, a *virtual rail* can be provided in the form of electrical lines or otherwise virtually created. If all motor vehicles in a certain environment communicate with each other, each knows where the other is and can act accordingly. Changing surrounding conditions can be communicated to the vehicles in a timely manner via the Internet or special sensors. Add to this better, integrated all-round cameras and many distance-measuring units, etc., and there is hardly a problem standing in the way of the self-driving car that cannot already be solved with today's means—as long as you do not drive off the road. Whether it will be accepted by customers is perhaps a financial risk at the moment, because many drivers still want to drive their cars themselves, even though computers could do it better. When the new technology becomes mandatory is a question of time that will depend on political will. Feelings, awareness are certainly not necessary for these computers in this case.

Consider one more specific area of automation: For decades, experts have been complaining that, for both economic and political reasons, they have not managed to bring the area of air traffic control completely into the digital world. Communication between aircraft pilots and ground stations still takes place via analog voice radio, which has been left behind in the mobile communications sector since the 1990s. It is only now that the air traffic sector is facing this huge transformation. Everything is to be digitalized and computerized, which will not only significantly increase safety, but also flight density. This goes so far that the voices are getting louder that are calling for doing away with pilots altogether, because humans constitute a major risk in many respects. The basic idea is to no longer determine flight paths via highways of the air, but to let the computers calculate the most energy-efficient flight path or to base it on other optimization goals (speed, safety, etc.). In this way path changes could be made quickly at any time, for example in the event of weather changes, without the need for cumbersome queries—pilot calling ground station. This means that the routes are precisely calculated by computers. Wishes of the pilot, perhaps to arrive a little earlier or to fly over more beautiful areas, no longer play a role. The machine scrupulously ensures that all criteria such as safety, energy savings, etc. are met without exception. In this case, the intelligence of the machines is limited to searching out specific information in databases and calculating specifications according to predefined algorithms. From today's perspective, all solution steps required for this can be accomplished with the help of classic automation solutions, which simplifies questions regarding safety. Human intelligence based on feelings and awareness for machines is certainly not necessary here either.

And this brings us to the central question: What cannot be done with classical automation systems, classical artificial intelligence (AI and today's (technical) cognitive science (CS)? When are the methods used with them (deep learning or tools from companies like Deep Mind) no longer sufficient? How can we imagine the dividing line between the two different application domains? When does the machine need feelings and consciousness? Does it need them at all?

If the answer is affirmative, this leads directly to the next question: What methods are needed for systems that require feelings and consciousness? What are the consequences? Are the consequences even manageable, or are we not behaving like *The Sorcerer's Apprentice*, who eventually lost command of the situation? What are feelings and consciousness anyway? Many claim without hesitation: These are the human characteristics that distinguish us from machines. Really? This general statement alone must make you wonder, even if it has been and still is expressed by great thinkers and philosophers. Philosophers have been wrong before. I would like to remind of news like: "Biologists create artificial life" (Forschung & Wissen, 2014) or "Synthetic Genome Brings New Life to Bacterium" (Science 2010; Powell 2018). 100 years ago, it was not even thought about that scientists could create life from dead matter.

Or bring the following thought to its conclusion: Once one is able to simulate the brain, one can also emulate it. This means that this simulation construct of the brain—the functional structure of the human brain—can be incorporated into a robot. There is no technical aspect that would speak against it. How then does this consciousness, these feelings of the robot differ from those of humans? *They cannot be human consciousness and feelings, because we are made of flesh and blood, while the robot of sheet metal, plastic and a lot of semiconductors.* What emotions and feelings will it have? Can it even be considered to have feelings? Here doubts and contradictions must arise and must be talked through.

I have set myself the goal of answering these questions with natural scientific methods. Why is it so difficult to answer them in *brain science* and *technology*? For me, the following central questions did become apparent: *What tools have brain researchers lacked to be able to clearly answer these questions?* Why has it not been possible to engage more intensively with brain research in the field of AI and (technical) cognitive science? There is probably no doubt that AI and its derivatives have worked very hard and have been able to achieve great things. Think of developments that are now widely used in industrial products. Such are cited, for example, in Deutsch (2011), Dönz (2015), Lang (2010), Muchitsch (2013), Palensky (2008). In business management, these are especially knowledge-based systems, rule-based expert systems, business rule management systems, or in automation neural networks, decision trees, and fuzzy logic (Hawkins 2004, pp. 12–13).

Excellent successes have been achieved in all of these areas. Our Internet is simply inconceivable without AI. Think also of talking robots, chess computers, pattern analysis, high-speed trading, programs from Google or Amazon, filter bubbles, ..., the list could be continued endlessly. *But a robot still cannot serve me tea or help me get into my coat.* I have never lost sight of this proposition in my life. In the end, it forced me to deal intensively with neurobiology, with psychology, but above all with psychoanalysis. And this has its deeper reason.

As a professor for chip design, especially for the development of microprocessors and other smart devices, I taught a crucial theorem in this field for over 30 years: Smart devices must always be designed from top to down. This means that first the application (today the term *app* is on everyone's lips), i.e. its use, must be defined. The application dictates the requirements for the processor, i.e. what the function of the processor must accomplish. This is the "top" in the top-down design. This function can then be broken down into sub-functions and those into sub-sub-functions and so on until you finally arrive at the transistor, an extremely small sub-sub-...-unit (or: -function). What does this mean for humans? What is the top of the process: "human being"? Clearly it must be the psyche. So, before dealing with neurons and sensory cells (receptors), one must first model the psyche. This brings psychology and psychoanalysis into play and, as will later become apparent, especially psychoanalysis. At this point one question arises as a pivotal question:

How to describe the relationship between the nervous system and the psyche?

This is supposed to be the central question of the whole present book, which has to be answered.

Of course, there are many inspiring books that have enabled me to answer all these questions. And it should be self-evident, we technicians alone cannot be the experts for this subject. The experts can and must come especially from neuroscience, neurology and psychoanalysis.[2] There are fundamental works which I have always recommended to my students. They have enabled me to follow the *natural scientific path* consistently for more than 20 years now: *The Feeling of What Happens* (Damasio 1999), *The Brain and the Inner World* (Solms and Turnbull 2002), and *The Brain in Action, Introduction to Neuropsychology* (Luria 1973). The reader will notice that much knowledge from these books has been used as a basis for

---

[2] Why particularly from psychoanalysis and why psychology is not sufficient, will be explained in detail.

our project, and—crucial for us researchers—that our *technical simulations* and *natural scientific experiments* so far do not show any decisive contradictions to their findings, although most of the results of psychoanalytic research could not be tested *natural scientifically* via experiments until today, if one leaves aside statistical methods. But first, before this can be explained, one has to formulate and discuss, like in every *natural scientific* experiment, basic assumptions and boundary conditions.

Computer technology—my professional field—requires that you constantly work your way into new areas, as computers are more and more integrated into all kinds of devices, systems and processes. This requires—due to the necessary electronic adaptation—to have close look at the new areas and to analyze in a differentiated manner. As a result, one often develops new models in cooperation with the relevant experts. From the point of view of natural science and technology, one is in this way always entering uncharted territory. In order to properly (i.e. axiomatically) describe this new territory, more and more new terms have to be introduced. At this point I would like to do this, too. I do not want to permanently run into the embarrassment of having to name this system in a laborious way. So far, it has not been possible to give a name to the unity, which consists on the one hand of the part that is to be described neurologically, i.e. the neurological hardware, and on the other hand of the part that is to be described in accordance with the methods of information theory, of which the psyche is a part. I make here this seemingly tremendous step and call this entity from now on the Ψ-organ (psi-organ). The letter Ψ had been used by Sigmund Freud when he first made the attempt to account for the mental phenomena in a way that took also into consideration the conviction that they are somehow the result of the functioning of the nervous system.[3] Ψ is therefore supposed to stand for psychoanalysis. The term *organ* refers to the fact that the whole is to be seen as a unity, just like other organs in our body, i.e. parts of the human body to which a common function is assigned. It is intended to express that the term Ψ-organ is considered as the unity of the neurologically described functional units such as the neurons, the neuronally connected sensors and actuators on the one hand, and, on the other hand, the informationally described functions of the psyche (ego, id, superego, etc.).[4]

---

[3] This first attempt failed, but it led Sigmund Freud to conceive the theory of psychoanalysis.

[4] Since the term brain refers to the neuronal part inside the skull, it forms only a certain neuronal part of the Ψ-organ and does not refer to the psyche.

Chapter 2
Philosophy, Consciousness and Axiomatics

As long as I was offering topics for bachelor and diploma theses and dissertations, there were always students who found the topic of artificial intelligence exciting, but then demanded to contribute their ideas independently of psychoanalysis. They could not identify with psychoanalysis. Could I accept that? Would this not cause us in the team to lose sight of the goal and thus lose time? At first, I had my doubts, because I was—and am more than ever—convinced of my approach, but I agreed, because young people can develop an enormous potential, if one gives them the necessary freedom. And at the end of the day, it is the results of the experiments that decide what is useful and less useful, not our feelings. Nevertheless, the discussions and arguments ultimately help us to work out the right argumentation for our methods and principles. Unfortunately, over time I had to realize that the problem has to be seen more comprehensively. If one has not yet been able to develop the relevant expert knowledge, *one is easily tempted by simplistic solutions.* Which is understandable as in this way one does not have to laboriously acquire additional knowledge. Hence, there is a lot of literature about AI and psychology, which describe human behavioral principles. If a well-founded mathematical background is added, nothing seems to speak against simple solution schemes. Based on them one can in most cases design programs without having to work out fundamentally new knowledge. And in an unknown area, pitfalls are hard to spot. The artificial agents modeled based on this approach often show an astonishing result and perhaps even an apparently human behavior—at least as long as one does not understand the term "human behavior" in the deeper sense, i.e. does not question it more closely.

Some students who contact me want to use our tools to directly design programs for the gaming industry, which promises a lot of money. Others, more scientifically interested students, harbor the desire to work on intelligent models that promise great success, such as the Blue Brain project, in which IBM and support funds have already invested a lot of money to date. At the same time colleagues from Japan, Taiwan, India, Canada, South Africa, etc., approached our team to join forces to develop robots with more "human intelligence". But what constitutes "human intelligence"? I was not convinced by the proposed approaches. I have spent my

D. Dietrich, V. Hartmann Cardelle, *Simulating the Mind II*,
https://doi.org/10.1007/978-3-031-69530-8_2

life doing applied development and research for the aerospace, automotive, facili-
ties- and communications industry, as well as for automation in general, i.e. for
products that ultimately had to be sold. In contrast, in this case I am concerned with
solving a fundamental question of natural science; I am concerned here for the first
time with pure foundational research—at least as far as my personal work is
concerned. In doing so, I have always had the feeling that the path via classical
artificial intelligence cannot solve my central problem:[1] I want to establish the
connection between the physiological described part of the nervous system (resp.
of the neuronal system) and the psyche in a scientifically verified way, i.e. to model
the Ψ-organ as a unity of neurons and psyche—and not just to look at certain
phenomena of human behavior or the behavior within interpersonal relationships.
What must a consistent model look like that I can then integrate into robots? How do
these robots feel? What is human? What is intelligence? Which functions in the
Ψ-organ are responsible for it? What distinguishes humans from today's machines?
Why cannot a computer system monitor an airport without human guidance? Why
do humans still have to sit behind the monitors and make the final decision to act?
Why can a robot still not serve me coffee like a waiter or waitress? What is the
central problem? *How do I have to approach these central questions?*

I discussed these questions with students in seminars over decades and had to
realize with great regret that not only I needed the long, laborious way to come to my
present understanding in the SiMA project[2] (which is the basis of this book). It takes
some time for everyone to think through the relevant contexts that allow the
neurological and psychological parts to be linked in a consistent way. I had to give
each student the time they need to get there.

In this way, I finally came to the conclusion that it is futile to talk about project
SiMA with colleagues within my faculty, since they were all specialists in other
subjects—which is the essence of a faculty. They live, as it were, in a different world
of thought. They do not want to deal with questions of the psyche, even less so with
psychoanalysis, which for many is—as I had to hear again and again—an "obsolete
science".

Psychoanalysts, on the other hand, usually had the problem of not wanting to
bother with *theories from natural science*, especially not mathematical ones, since
they would not be able (or maybe: wishing?) to grasp it anyway. So most of them
found a good enough argumentation for not having to bother with the subject at all.
But such prejudices are an excuse for not thinking outside the box and—over time—
they make you unable to do so.

This brings me back to the problem many of my computer engineering students
were facing: For the most part, they first had to be convinced that the description of
simple behavioral phenomena is not sufficient to create AI that can generally act in a

---

[1] Anyone familiar with Jeff Hawkins' book (Hawkins 2004) will find that I am not alone on this
point (Hawkins 2004, p. 16). His analysis largely coincides with mine in this respect, only that my
interpretations and conclusions ultimately aim in a different direction, which I will come to later.

[2] SiMA stands for: Simulating the Mind and Applications.

"human" way. Everyone has to work out a mental background that provides the basis for the new path. How does this basis look like? Baruch de Spinoza (Auerbach 2017) would say, that first the concepts must be worked out and defined, that is, an axiomatics must be provided for the task of research, because otherwise you get an empty rhetoric with which you can prove everything and nothing. Like any natural scientific theory, it has to rest on conceptual, solid (natural) scientific pillars. The philosophical pillars that are decisive here will be explained in the rest of this chapter. The other pillars, the purely scientific ones, will be discussed in the following chapters.

2.1 Monism

Baruch de Spinoza wrote in his work *Ethics* (1765), mutatis mutandis: "Man created God in his own image ..." and set himself with these fundamental considerations against the zeitgeist and lighted a beacon of enlightenment in our western hemisphere. It was important to him to expose contradictions in philosophical and religious statements (Mauthner 2015, pos. 410) instead of blindly believing them. He became a rabbi at a young age, which can be seen as an indication of his enormous logical abilities and intellect. However, he was not only interested in the Jewish world, he also worked his way into the literature of the Christians and was expelled from his Jewish community. He found friends among the Protestants, who, after a certain time, also distanced themselves from him because of his radical criticism of not simply accepting the faith, but constantly questioning it. He did not fare any better on the Catholic side (Yalom 2012, pp. IX, 318).

In the religious works he read, he repeatedly encountered contradictions. He preferred, however, to leave questions unanswered rather than simply believing what he did not understand. His conclusion was that he defined God as infinite: God is everything. This monistic way of thinking—his pantheism—impressed many personalities, like e.g. Gottfried-Wilhelm Leibniz, Gotthold Ephraim Lessing, Heinrich Heine, Isaac Newton, Johann Wolfgang von Goethe, Albert Einstein, Norbert Wiener, and others. They all emphasized Baruch de Spinoza's great achievement. This gives natural science a clear foundation (Yalom 2012, pp. IX, 36, 38; Restetzki 2019; Auerbach, B., 2017, pos. 238; Digressions&Impressions; Oxford Handbooks online). The fundamental tenet of his thinking was: Relationships must be defined axiomatically—that is, unambiguously and clearly. This shall be explained in more detail in Sect. 2.3. The mathematical way of thinking was important to him and implied for him to treat natural laws as a given. As such he saw laws of natural science provable.[3]

---

[3]Two notes have to be added: The first is that the term *provable* is to be understood in terms of Baruch de Spinoza's zeitgeist. From the point of view of today's science only purely cognitively developed derivations are provable, e.g. those of mathematics. Laws of natural science are not

If one transfers these considerations to the present time, a simplifying consequence can be drawn for the natural scientist: It is necessary to distinguish different areas of science. One area is natural science. Everything that cannot be grasped axiomatically and thus by natural science must be left open for the time being. What about an immortal soul in human beings? What happens to it after death? And where does it come from? All these are questions that cannot be dealt with by natural science. These questions must be taken up by philosophy, the humanities or religion and theology. In the present text, however, it is assumed that all human behavior can be explained in the long run by natural science. The existence of mystical phenomena is therefore negated.

In so doing we commit ourselves to monism. In particular to the assumption that humans[4] have on the one hand a body, in the physical and chemical sense, and on the other hand a psyche; and that this body and psyche are inseparably connected with each other to one entity. Spirit, intelligence, communication etc. are terms which concern the world of the *psyche* and *information*, and this world then is based on the body (matter). The theory of information theory which describes the world of the *information*, is to be seen therefore as a different *method of description* in contrast to the *physical and chemical method of description*. *Energy* is a physical concept that thus defined and connected to physical methods of description. "Psychic energy", as it is assumed in psychoanalysis, is something completely different, that must not be associated with energy (i.e. the term of physics). For this reason, in this book the term *psychic intensity* (a term that Sigmund Freud himself used in *The Interpretation of Dreams* synonymously with "psychic energy" (Freud 2017, pp. 420–422, 447, 452–453) is used instead and will be discussed in detail later.

With this stance we distinguish between *abstract objects*, which are described purely on the basis of physical and chemical laws, and abstract objects, which exclusively process information and must be described by the laws of information theory. Accordingly, this information-processing organ must in principle have a hardware, that is described with physical and chemical methods and laws, and an information-technical part, which can only be described and explained with its own language (i.e. own set of methods and laws).[5] Information without body is a human illusion according to Baruch de Spinoza's statement (freely quoting) (Auerbach 2016, pos. 4048, p. 194). "The mind can envisage or remember past things only as long as the body lasts. Proof: the mind ...". In humans, the body of the information

---

provable. They can, however, be evaluated. I will come back to this basic distinction again. The second note is that Baruch de Spinoza, did not draw the conclusion that the Vienna Circle drew from this in their reflections on objectivism: An axiomatic clarity of concepts does not necessitate objectivity.

[4] The authors are aware that therefore highly developed animals also have a mental apparatus, which cannot be discussed here.

[5] This statement might seem trivial. Yet, it not only is a core statement in the SiMA project, but the crucial point to understand the core problem in SiMA. Many do not comprehend the consequences of this statement and therefore have massive difficulties when trying to understand the Mealy principle of computer technology, which shall be described in Sect. 5.2.

system consists primarily of the nervous system, or brain, and of the hormonal system.[6] The nervous system (hardware) consists of about 10 billion neurons with about 10 trillion synapses, although some neurons may have as many as 6000 synapses (Damasio 1994, p. 29–30). The mental apparatus represents the mental, which cannot be explained by physical or chemical laws, which is why the use of physical analogies when describing the mental apparatus not only can be wrong but often is heavily misleading (the prime example for such a misleading use of concepts is to associate "psychic energy" with physical energy).[7]

In the same way, it is nonsense to apply laws of the information world to the realm of physics and chemistry. I do not know how this should work. Both description methods must be kept strictly apart (as if they described two different objects), although they ultimately describe one and the same object. The problem about this object is that it cannot be described with *only one* set of methods and laws. This assertion might at first be difficult to "grasp". It will be discussed continuously in the following, by which it should become more and more comprehensible. Yet I need to emphasize once again: In the course of my work as a supervisor of many dissertations, I had to realize that almost no scientist grasped this point with ease. Many had a very hard time getting to this point, even if they were "only" dealing with topics of technical electronics like the chip design.[8] Scientists in computer technology, however, had recognized relatively early on that the physical part of the computer is subject to different laws than the information part,—a differentiation that Sigmund Freud had made before too.

Let me briefly sketch it out: John von Neumann developed the classical information theory based on statistical methods (Shannon 1948; Zemanek 2001). George H. Mealy has developed a model that unifies both the world of information with the physical world. Siegfried Wendt (2016) has elaborated this in detail. In the standardization committee, the ISO/OSI model—the layer model—was developed with a great deal of effort, which defines the hardware and six information layers (Walke 1987), i.e. combines both views. So, the technical information theory of today is no longer limited to Claude E. Shannon's classic treatise, but also includes many other scientific insights.

---

[6]The hormonal system is also part of the human information system and must also be divided into the purely physical part and the information part in the same way. Yet it has not yet been incorporated into the model of SiMA, for reasons of cost and effort.

[7]Consider this example: In [Freud GW I, p. 561] Sigmund Freud draws the comparison between electric charge and psychic excitation. However, this comparison easily leads to an erroneous track of thought. Electric charge is a physical quantity whose spread can be described physically. Psychic excitation, however, represents the psychic valuation of information, as shall be explained in more detail. It is therefore itself an information. Since information is not a physical quantity it therefore cannot spread physically.

[8]A decisive reason why my young scientists struggled with the separation between the two worlds of description was often the attempt to "smuggle" analogies from the physical world into the world of information, which could only lead to confusion. As it happened in the case of the term "psychic energy".

In our technological world, we find the computer in many devices such as the mobile phone, the washing machine, the motor vehicle, the television set, airplanes and so on. This must lead to the all-important question: What speaks against applying the models developed by technical information theory and their thoroughly elaborated natural scientific methods and laws to the Ψ-organ as well? After all they are scientific methods and laws that aspire to general validity and thus are independent of the concrete applications. Or asked the other way round: Should there really be two different kinds of information laws, one for living beings and one for machines? Then there should be a scientifically founded distinction between these laws. Yet, I do not know it. And after the first artificial, living cells have been developed (Dickerson 1979; Deamer 2005; Buddingh and Hest 2017), this for me far-fetched thought moves very much into the mystical.

I am aware that this theory is so difficult to grasp for the simple reason that our upbringing has essentially always aimed to separate the body and the part described in terms of information. This becomes particularly difficult when, as it is common in religion, soul and psyche are associated with each other and the soul is defined as being immortal. If this is hammered into a child, it becomes difficult to break away from it. Even more so if later on, in the wake of academic education, the separation of body and the part described in means of information terminology is repeated in form of the separation of natural sciences and the humanities and loaded with jealous suspicion for the other academic field. Baruch de Spinoza's path of expulsions describes this all too well (Auerbach 2017; Yalom 2012). We owe him a great deal of credit for standing up against the mindset of the general public, even though he saw that his path could not have a happy ending. For humanity, his work and suffering paid off.

2.2 Needed Expertise

When I initiated the project with today's name SiMA around the year 1999, two fundamental questions were associated with it for me: How can we engineers succeed in making a machine think like a human being? And which experts can help us and in which way? A brief anecdote on this subject. An excellent student who was interested in the topic of AI explained to me his visions of it, which amounted to purely mathematical algorithms. I, however, explained to him that I thought this was the wrong way to go, because the field of AI had and has not come close to exploring the explanation that is important to me: the bridge between the hardware and the psychic apparatus. What went wrong? One point was clear to me: We technicians would have to bring in experts such as neurologists, psychologists and psychoanalysts. He reacted completely aghast, as for him the only acceptable scientists in this field were philosophers. I experienced such reactions again and again later at conferences, workshops, seminars and other events, where the philosopher was sometimes replaced by the neuroscientist.

It became clear to me that the answers to my questions about the methodology and way of the approach as well as which experts should be consulted would not only significantly influence the direction of the research, but also provoke discussions with persons of a different opinion—this all the more since I was looking for a really new approach.[9] Consequently, the answers had to be thoroughly worked out. We had to overcome the comfort of our habits that were shaped by the natural sciences. The mainstream should not be allowed to set us off course. The fact that these sentences are much easier to formulate than to consistently follow their spirit was not clear to us at the time. Everyone who approaches the subject for the first time in SiMA has to make the same experience. I noticed it again and again when students indicated that they wanted to participate in the project.

But let us start with the results achieved so far. AI developed outstanding algorithms, without which our modern society would be unthinkable today (Palensky 2008). For instance, the internet is impossible without AI systems such as knowledge-based systems, neural networks, rule-based systems, business rule management systems, fuzzy logic, deep learning, generative intelligence, BBAI (behavior based AI) and others like tools from companies like Deep Mind, have deeply penetrated our social, industrial and business world. Without their algorithms, many things would simply stop working. Considerable amounts of money are being made with them, and that is to be welcomed. However, the goal of understanding human thinking and thus being able to manufacture machines with the ability to think like a human has not really been achieved. The link between the *neurologically described* part and the *psyche* was first achieved and modeled in the SiMA project. But what are the reasons for this? Where are the pitfalls? The major figures in AI—as well as those in the world of robotics—kept predicting that we were on the verge of a breakthrough (Irrgang 2005; Hawkins 2004, p. 12), and yet, after a while, this wish turned out to be much more difficult to achieve than anticipated. What was and is the reason that previous AI and robotics have had such massive problems conceptualizing and, above all, modeling the bridge between the hardware (neuronal system) and the psychic apparatus?

Crucial for me is that I only have heard predictions of an AI breakthrough from computer scientists, electrical engineers, physicists, or mathematicians (Honey 2016, Part 2; Hawkins 2004, pp. 12–13; Burghart et al. 2007), who by virtue of their training cannot have the needed academic expertise, i.e. the knowledge, empirical experience and practical skills needed to be able to work out in a sound manner how the Ψ-organ and thus how humans "work".

---

[9]With regard to foundational research, it is difficult in today's academic landscape to take a path outside the mainstream, as research proposals have to go through the peer-reviewing process and thus have little chance of being approved. The only way out is often self-financing, which was largely the case with SiMA.

In order to understand complex[10] relationships in a new and unfamiliar world, you need a certain amount of time to become acquainted with the field. Reading up on the subject is not enough. One must have studied it, live and work in it, in order to develop the "feeling" for the intricate details of a discipline. All students of electrical engineering know the problems of having to familiarize oneself with the world of Maxwell's equations. But even after having understood them well enough to pass the corresponding exams, one is certainly not yet an expert in them. The same applies to the laws of quantum theory. And yet there are numerous biologists, philosophers, physicians who apply laws in their chains of argumentation that cannot be assigned to their field of expertise. For example, John C. Eccles, an outstanding researcher and a personality in the field of neurology, connected quantum theory with consciousness (Eccles 1994, p. 145), although he certainly can never have dealt with quantum theory in depth. In the same way, there are mathematicians, computer scientists, electrical engineers who explain what intelligence is, what feeling is, or what consciousness is (Blum et al. 2005; Sloman 2009; Palensky 2008, p. 1), without having any academic expertise on the subject matter. I consider this as a kind of hubris or at least a failure to recognize the scope of the scientific problem—which is even more problematic. than a lack of respect for scientists from other fields.

How was the training of a psychoanalyst organized in Vienna just a few years ago? (I will explain later why I chose the psychoanalyst and not another profession for our SiMA project.)[11] The precondition was a degree in psychology, sociology, medicine or a similar field. First you had to complete the psychotherapeutic preparatory course and only then did the proper training begin. And in parallel to the theoretical studies, one also has to lay "on the couch" oneself, i.e. to undergo a high-frequency psychoanalysis in order to experience and thus better understand what unconscious processes are, which are my own conflicts and problems and which are those of others. This can only be done by reflecting and working through with help of a teaching analyst one's own transference and counter-transference, never through pure self-observation and even less via pure logical reasoning. I do not have exact data, but from my circle of acquaintances it is always an average of 7–10 years until the whole training program for psychoanalysis could be considered completed (the years of prerequisite study are not taken into account). If I look at the education of an engineer (for example, computer technology, my own degree), the average was about 6 years in total. And after the studies it was still necessary to gather years of experience to get a feeling for the specifics of the subject, on the side of psychoanalysis the clinical experience, on the side of engineers the industrial experience. All of this takes time, a lot of time. If you add up all the years of

---

[10]By a complex relationship it is meant that one cannot achieve a complete description of a process, because information is missing for this. In contrast to this, one can describe for example a not complex behavior completely in all variants. This is elaborated in more detail in Sect. 4.1.

[11]New and alternate methods of training new psychoanalysts and their consequences are not relevant for the aim of the book. What I want to make understandable to the non-psychoanalytic reader is that the training of psychoanalysis demands an enormous learning effort.

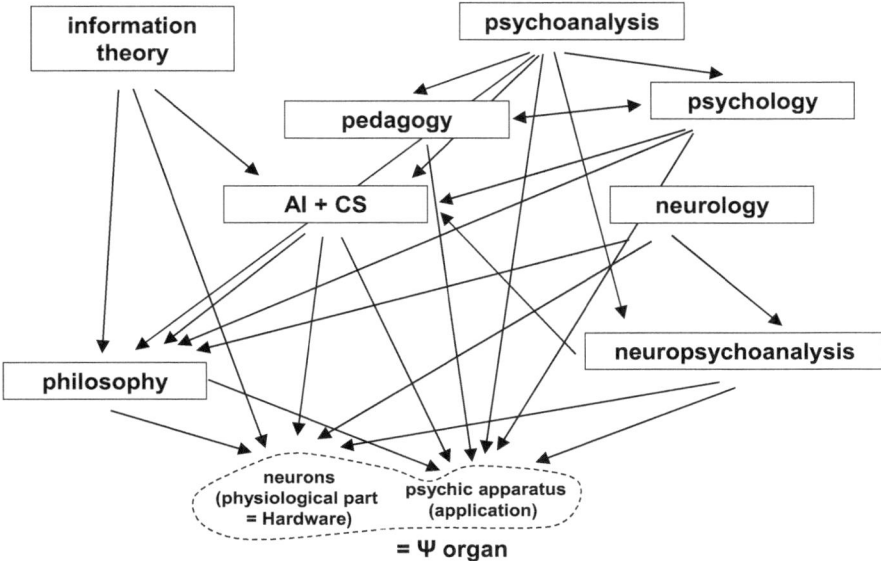

Fig. 2.1 Authoritative sciences dealing with the Ψ-organ. *AI* artificial intelligence, *CS* (technical) cognitive science

education and training, you can easily arrive for psychoanalysts at 11–14 years and for engineers up to 7–8 years.

So, if one wanted to acquire the knowledge of both worlds, one would obtain a training time that is neither feasible nor reasonable. The engineer can only roughly acquire psychoanalytic knowledge, but the really deep understanding of this world will have to remain largely inaccessible to him. The same applies vice versa to psychoanalysts, as already explained above (Dokaupil 2006a).

What is the consequence? That only cooperation between various scientific disciplines can lead us to the goal of acquiring knowledge about the Ψ-organ and, in particular, its structure. But which disciplines should these be?

In Fig. 2.1 we have a graphic showing the academic disciplines that deal in some way or other the Ψ-organ as well as their inter-relationships. We start at the top left: It should be clear by now that engineers of technical information theory, my field, do not have the training and the necessary know-how to explain the functions of the Ψ-organ. The same holds true for philosophers. However, they can critically examine a functional model of the Ψ-organ as well as the methodology for its development to see whether logical inconsistencies have crept in. Pedagogues have a different goal than to acquire a model of the Ψ-organ. They care about questions and methods of education, for which they develop models of human behavior. Some of them use the results of neuroscience, which unsurprisingly often leads to heated debates. Huber (2012) as well as Hasler (2012) point out with regard to this matter that the supposed connection between neuroscientific and educational models has so far never been scientifically founded in any way and did certainly not meet the standards of natural sciences.[12] Psychologists primarily study certain behavioral phenomena, but have no functional model that computer engineers could map into a scientific, functional model. Neuroscientists, on the other hand, try to work out a clear scientific, functional model of the nervous system. The same is true for psychoanalysis. But in contrast to neuroscientists, who focus on the hardware of the Ψ-organ, psychoanalysts have developed a basis for functional modeling of the information system of the Ψ-organ: the second topical model (also called: the structural model), which corresponds exactly to what a computer engineer would have in mind. This will be explained in detail in Chap. 6, when the preconditions have been met.

As a consequence, some neuroscientists and psychoanalysts joined forces in 1999 and founded the NPSA (Neuropsychoanalysis Society) to work out a bridge among all the various disciplines, that means the central question of this book.

It was therefore only logical that engineers and scientists from AI and (technical) cognitive science joined forces with scientists from the NPSA (Dietrich et al. 2017b; Yovell 2009, p. 251). Since, as it has been mentioned and shall be explained in more detail, modeling according to computational theory must strictly adhere to the top-down approach, the "top" has to be modeled first, i.e. the psyche (which I call in the following the psychic apparatus[13]), before the details of the "down", i.e. the hardware, can be developed. The collaboration with psychoanalysts must therefore be prioritized, but of course the neurological aspects must not be disregarded. The details of the model must not contradict neurological principles, which is why a neurological voice was always present in the SiMA project, along with scientists from psychoanalysis.

But this already shows a fundamental problem of the SiMA project: Many psychoanalysts see their science in the field of the humanities (i.e. as a soft-science) (Harlfinger 2007), which clashes with the methods of natural science. And they are even supported in this opinion by philosophers like Karl Popper or Jürgen Habermas. However, if one scrutinizes their argumentation and use of terminology on this question, it leaves much to be desired (Doblhammer et al. 2015; Hartmann Cardelle and Dietrich 2022). This leads directly to the next chapter.

---

[12] Since neuroscience sees itself as a natural science, but pedagogy usually sees itself as part of the social sciences, therefore a natural scientific foundation must be worked out for the connection of both fields, if natural scientists are to be convinced.

[13] As I will explain in more detail, psychoanalysis developed a functional model which, from the point of view of information theory, can be axiomatically reproduced and thus tested: the so-called psychic apparatus. I will therefore henceforth use this term instead of others like psyche.

2.3 Resistances and the Open Problem of Axiomatics

It is crucial to call a spade a spade. We in the SiMA project have always had to fight a fierce headwind. From many sides. And although more than 200 scientific publications have already been published internationally by SiMA collaborators, similar questions and discussions arise after presentations. Are they exclusively motivated by prejudice? Entrenched prejudices prevent discourse, even if they are proven wrong by our experimental results. Let us start with the decisive resistance in psychoanalysis.

It is said that humans think differently than machines. The brain is not a computer (Leuzinger-Bohleber 2008, p. 39; Dietrich et al. 2012), is the common wisdom. How does one come up with this frequently formulated statement that is usually seen as a truism? What is its scientific justification? Or is it rather motivated by wanting to avoid the fourth blow that humans have to undergo? Sigmund Freud pointed out three basic narcissistic blows: the cosmological, the biological and the psychological blow.[14] Is there, then, a further blow, prompted by the information theory of computer technology, on the basis of which "thinking" machines can be created? Or is there after all really an information theory of the Ψ-organ, which is independent of the information theory of the computer technology?

Let us try on both sides, that of engineering and that of the sciences within neuropsychoanalysis, to forget the prejudices for a moment and to acknowledge the other side's knowledge. One will then understand that all fields in natural science are based on the same physics, chemistry, biology and also on the same information theory, including for the understanding of the nervous system or the Ψ-organ. Furthermore, one will have to recognize that physics and chemistry obey different laws and rely on different presuppositions than information theory, from which follows that at least two theoretical fields are necessary for understanding biological entities: physics and chemistry, on the one hand, and information theory, on the other hand, which needs to be specified in more detail. And one will have to recognize that the Ψ-organ can only be modeled and thus simulated in a holistic manner, if one grants the required space to both fields of science, which, however, must not contradict each other. Exactly this aspect shall be addressed here.

The resistance from the engineering community is formulated more aggressively. Team members were told at conferences that their work is good and interesting, but why did we base the project on the outdated science of psychoanalysis? After all, Sigmund Freud's theories are no longer appropriate. Why should one bring together

---

[14] Cosmological blow by Copernicus: the earth is not in the center of the world. Biological blow by Darwin: humans are nothing else and nothing better than animals. Psychological blow by Sigmund Freud: human consciousness is based on unconscious processes. The blows for which Freud coined the saying that the ego is not master in its own house (Freud 2020c, p. 30; Schülein 1999, p. 37).

such an outdated theory from the humanities[15] with modern natural science? Or even harder, as it was formulated to us several times: Technology and esotericism or mysticism do not belong together!

We grew tired of these attacks, and after carefully avoiding the term *psychoanalysis* in scientific and industrial funding applications, SiMA was successful in obtaining funding. Needless to say, what can be concluded from this.

But let us come back to the question posed above: Is it solely prejudice that prevents the scientists from natural sciences and the humanities/social sciences from collaborating? Let us take up the term "thinking" used above. How many conceptions of the term "thinking" might be out there? It follows from this consideration: As long as terms like thinking, feelings, emotions, psychic energy, perception, symbolization etc. are not clearly (axiomatically) defined, a fruitful discourse with technicians, let alone a technical realization, is not possible. The technician needs a clearly (axiomatically) defined terminology.[16]

This leads us to the central problem between natural science and humanities (given that one accepts this separation for the sake of simplicity). Natural sciences, and especially technological disciplines, need an axiomatic terminology, whereas the humanities decisively rely on hermeneutics. Even if the humanities would wish a clear axiomatic terminology, they would not be able to finance the necessary expenditure for the creation of such an axiomatic terminology. But one thing at a time.

Bertrand Russel, a British mathematician and philosopher, together with Alfred N. Whitehead, wrote a work of fundamental significance for natural scientists: Principia Mathematica (Russel and Whitehead 2009). The theory is based on the assumption that all elaborated models have two distinct components: Axioms and a set of rules. This assumption is independent of what the model describes, be it a technical object or a church. As an example, I would like to address neurology and psychoanalysis right away. To create *a unified model for neurology and psychoanalysis*, the *goal of the SiMA* project, thus implicitly necessitates the definitions of *axioms* such as *drive*, *drive content*, *emotion* or even *consciousness*, and to create the set of rules describing the relationships of these concepts. Natural scientific thinking nowadays presupposes such a basic framework.

Figure 2.2a is intended to illustrate the issue. With a larger number of terms, the network can quickly become very, very complicated. But only in this way the unambiguous relationship between all terms is possible. A simulation is only possible in this way.

---

[15] It should be explicitly pointed out here that I do not classify psychoanalytic theory as a pure discipline from the humanities (see for detailed argumentation (Hartmann Cardelle and Dietrich 2022)).

[16] This statement about the need of a clear axiomatics free from contradictions is often misunderstood by the psychoanalytic community. It is the terminology that must be unambiguous, but the data or information in a system, such as in a machine or in the Ψ-organ (especially in the psychic apparatus), may very well be contradictory and are so in complex systems.

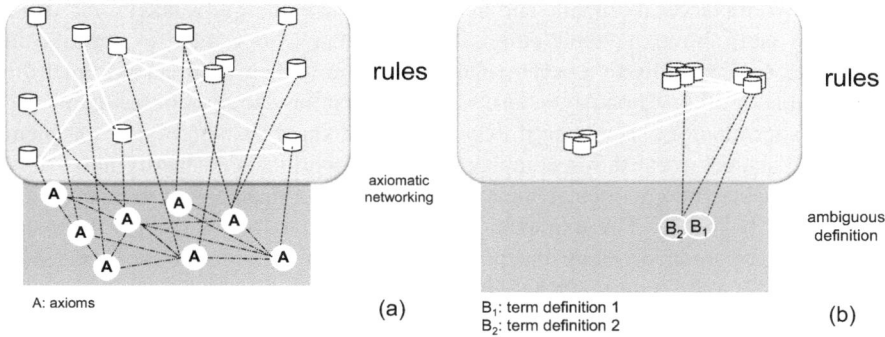

Fig. 2.2 Two ways of thinking: (**a**) on the basis of unambiguously defined axioms and (**b**) with *fuzzy* assignments of terms. B_1: *affect definition 1*; B_2: *affect definition 2*

For two reasons, the alarm bells of psychoanalysts reading this text are certainly ringing now. First, is not this a contradiction: psychoanalysis and unambiguous terminology? Just think of the first and the second topical models. How can they be brought together without contradiction? Second, how is one to work clinically in psychoanalysis on an axiomatic basis? If we were to introduce an axiomatic terminology would not this demand from the psychoanalysts that they teach the patients the *unambiguous* definitions of all terms before they can converse? Of course, this can only be nonsense and is not what is meant here. To make understandable what is actually meant, I shall elaborate the basic knowledge of technical information theory and its preconditions. Fortunately, this is not very difficult to understand and in the following chapters the explanations shall be given.

Let us first talk about the realm of the humanities to show that the goal of SiMA is not to dismiss the current way of thinking in psychoanalysis, on the contrary, it is absolutely necessary.

Psychoanalytic practice relies on the hermeneutic method. According to it, the meanings of the terms are used according to the context. Each author can vary the meaning of a word, related to the context,[17] in order to be able to convey their ideas as accurately as possible. Take Fig. 2.2b, as an example: There only one term is shown, but with two meanings, B_1 and B_2. This has serious consequences as the set of rules now no longer is consistent. With each further modified definition there are several parallel sets of rules. Let us examine the psychoanalytic term *affect*, for which there are various definitions in the literature. Such definitions that exist in parallel to one another must inevitably lead to contradictory contexts—i.e. to an inconsistent *set of rules, as I said before. This alone is a reason why there are several clinical methods of psychoanalysis.*

---

[17]This presupposes that the reader is able to think his way into the manner in which the author establishes the meaning of the words (hermeneutic presupposition).

The two methods, axiomatic and hermeneutic, have their advantages and disadvantages and, above all, their purpose. Any patient who wanted to describe his problems axiomatically to a psychoanalyst—assuming there were an axiomatically defined list of all psychoanalytic terms—would first have to familiarize themselves with this terminology. But even if they did, only a small fraction of their problems could be shared, because his complex emotional world would hardly allow for a comprehensive presentation of their problems in this manner. And vice versa it would not be better: The explanations and interpretations of the psychoanalyst would not be understood by the patient, because the patient would necessarily have, if at all, a different understanding of certain terms.

In contrast, natural sciences cannot work with the hermeneutic method. They need an unambiguous definition of terminology and interrelationships that is free of contradictions (Fig. 2.2a). No technical development would be possible without exact (axiomatic) definitions. We cannot have different definitions for terms like screw, hammer, electrical current, app, operating system, voltage, phone number or ID. For this reason, the industry invests enormous sums in standardization bodies to clearly define and also specify (= standardize) the necessary terminology. These are organizations like ISO, ITU on international level, CEN or CLC on European level, DIN in Germany and ASI in Austria,[18] to name just a few examples. This specification of the exact meaning of a term (and the simultaneous avoidance of synonyms) applies to all areas of natural science, because ultimately everything in natural science is broken down into functions and quantities of different fields: physics, chemistry, biology, neurology, information theory etc. For certain disciplines, standardization committees are not necessary for this purpose if there is no industry behind it that wants to produce products based on it. But as soon as something is produced, an axiomatic terminology must be developed so that every company knows how to design a product if it is to be interoperable with others. In non-manufacturing fields, such as mathematics, the scientists themselves have to make sure that there are no inconsistencies in the terminology, but this is relatively easy when mathematical formalisms are used.

If one wants to model and simulate the functions of the Ψ-organ, one needs a simulation program, which has to rely on a clear (axiomatic) terminological basis.

A crucial step in transforming the psychoanalytic theories into a natural scientific system is to develop a terminology that is unambiguous and without contradiction. *This is the greatest challenge in the SiMA project today*, because not only are there different schools within psychoanalysis, but, as just explained, it is not possible to approach this project in a hermeneutic manner. As a scientist, I have worked for a long time on precisely this issue in national and especially international committees. For example, before the introduction of computers, various professions, such as

---

[18] ISO: International Organization for Standardization, ITU: International Telecommunication Union, CEN: Comité Européen de Normalisation, CLC: European Committee for Electrotechnical Standardization, DIN: Deutsches Institut für Normung, ASI: Austrian Standards International (ÖNORM).

electrical technology, bricklaying, chimney sweeping, mechanical technology, carpentry, etc., were largely organized independently of one another and each had its own standardized terminology. However, the computer and the networking of the various controls, sensors and drives forced the organizations to sit down at *one* table with all the different experts. They had to work out a *common* language for their common concerns. And they had to acquire *one* understanding of the various trades with which their products would be interconnected. *At the beginning of this effort, all the involved experts threw their hands up in horror and declared that this was completely impossible.* There were many understandable reasons for that, after all behind each field were many different disciplines with their century-old traditions, completely different methods, theories and experiences. Should all of this be called into question?

This evolution of the terminological foundations began about 30 years ago and is still resonating in many areas. The national and international organizations spend each year a great deal of money on this for a good reason.[19] Therefore, it is not surprising if today psychoanalysts and neuroscientists also see an impossibility in this demand, already for financial reasons alone.

In SiMA, engineers, psychoanalysts, neurologists and psychologists have worked together on such an axiomatic terminology for more than 16 years. The result is an Excel file (Dietrich et al. 2015) with about 220 terms (axioms) in German and English. Figure 2.3, shows a short excerpt of it.

The challenge is that each time a new term (i.e. axiom) is added, it must be cross-checked with all the others to ensure that there are no contradictions between them. Any contradiction can lead to an incorrect set of rules (the functional system), which must inevitably lead to errors in the model's simulation experiments. In practice, this means that *a newly added term can lead to a modification of some previous definitions*, which in turn leads to a *modification of the previous model*. This is what makes the whole procedure so expensive in engineering, which is why people often hesitate when someone wants to introduce new terms. In the case of SiMA, this means that in spin-off projects this point must always be taken into account, both in terms of content and cost.

---

[19] Just to give you a rough idea: Such international meetings take place at least once a year and a meeting usually costs between $50,000 and $100,000. If you take into account the time for preparation and follow-up work of the experts, as well as the fact that there are commissions for a wide variety of areas, you can envisage the amount of money that the industry and governments spend on this task.

Expression	L	Definition	Remark
Quota of affect	2/ 3	The *quota of affect* is a type of *valuation* that originates from the *drive* and *drive tension* of *layer 1*. It can cathect *conscious* and *unconscious* contents and can be shifted by *unconscious mechanisms*.	When *quotas of affects* are discharged, they cause affects, which can be sensed *consciously*.
Drive	1	The *drive* attempts to adjust the *organism* towards an aim using a *tension condition*.	It is information of *layer 1*, which will be coded to a *drive representation* in the interface between *layer 1* and layer 2.
Drive content	2/ 3	The *drive content* is composed of the *source of the organ*, the *drive aim* and the *object of drive*.	It is a simplification using the expressions *drive content*, *aim of drive* etc. Because these values are part of *layer 2* and 3, the definitions should be *drive content representation*, *aim of drive representation* etc. However, because of the unambiguity given, the simplification can be accepted.
Object of drive	2/ 3	The *drive* can reach the aim via the *object of drive*, which can be a *person*, or an (partial) object of any type or a fantasized object.	See the remark of *drive content*. The *object of drive* is the most variable part of the *drive*, which makes it possible for the *drive* to reach its aim.

(a)

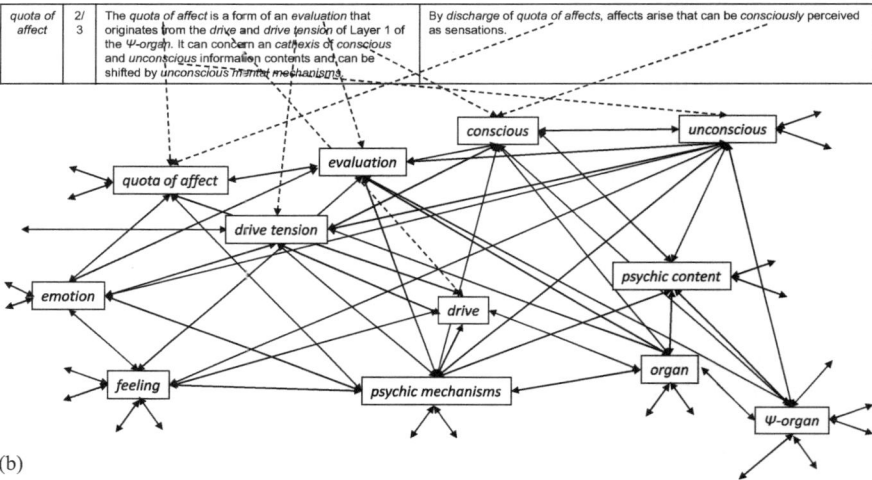

(b)

Fig. 2.3 Examples of term definitions in SiMA (axioms in SiMA). (**a**) In a table as text definitions, (**b**) dependencies of the terms. *L* layers 1–3 of the three-layer model discussed in Chaps. 5 and 6

2.4 Can Questions in the Humanities Be Answered in a Natural Scientific Way?

In my student days—the more I dealt with philosophy, church, cultural revolution, psychoanalysis, etc.—I came to the conclusion that I had to leave the church, since it appeared to me as a self-deception. So, I left the church. But then the question emerged: What could provide the moral foundation for a society when religious morality has become obsolete? The ideas of the Cultural Revolution? I studied their ideas for a long time through ideologically colored journals like the Beijing-Review. Over time, strong doubts arose, and today the cruel crimes of this ideology are becoming more and more visible (just read Wieser (2016), for example). This raises the question: Could such mental aberrations be prevented if one had a tool with which one could scientifically verify why such ideological and/or religious ideas fall on fertile ground in our Ψ-organ?[20] Or a simpler example, and formulated in a more general way: Would it not be politically and economically interesting if psychological or sociological relationships were testable on a natural scientific basis? It would be a great evolutionary step if such a reliable tool were generally available. And exactly this aspiration is made in the SiMA project: Ideological or religious conceptions could be simulated for social effects with such a tool by applying individual psychodynamic processes to a mass and its group dynamics. Whether one simulates 1, 1000 or 1,000,000 of Ψ-organs is mostly a question of effort. On the basis of these considerations, one could assign different individual characters to 1000, 100,000 or 1,000,000 agents and thus conduct sociological experiments. These experiments would go far beyond today's behavioral models, and the quality would be far beyond today's statistical methods in sociology. Thus, large-scale sociological studies with simulation experiments would be conceivable.

In other words: If the model of the Ψ-organ can be evaluated (nota bene: not proven, this is only possible in logics or on mathematics) by natural scientific methods, then psychological and sociological questions can of course also be evaluated, within certain limitations, by natural scientific experiments. And these limitations are principally defined by the kind of the underlying experiments and the level of detail of the programming of the Ψ-organ models.

What am I aiming at? What are the decisive aspects in this context?

Firstly: at the beginning the remark was made that our Ψ-organ is the control system that manages all our thoughts, desires, decisions and thus is also at least partly responsible for all of our actions. Secondly: in the subchapter on axiomatics, on the basis of the considerations and conclusions of Bertrand Russel, his considerations were adopted that all scientific models and methods must be based on a suitable axiomatic terminology. This must be worked out in general for the modeling of the Ψ-organ. And thirdly: if the corresponding experts (neuroscientists, psychoanalysts, technicians, etc.) manage to develop a functional model of the Ψ-organ from the knowledge acquired so far, *then psychological and consequently sociological questions should be able to be verified scientifically.*

In what follows it is therefore important to be able to make a clear statement about this.

---

[20] I am not talking about logical, philosophical examinations, still less about religious argumentation. Philosophers have been wrong again and again. They always judge on the basis of their zeitgeist, i.e. a limited knowledge, which is why their conclusions have to be taken with caution. In contrast to natural scientific results, they can already be outdated tomorrow. Religious verification methods are completely outside of natural science. With laws of natural science it is different. Thus the trigonometric equations will hardly have to be improved. The same applies to Newton's law of force. However, laws and findings developed in natural science can be refined, when viewed in a less abstract way. I also do not count among the tasks of philosophers that they evaluate hypotheses and statements on the basis of physical and chemical experiments, that is the task of natural scientists.

Chapter 3
Neurological and Psychoanalytical Models

The goal is to develop a model of the Ψ-organ using natural scientific methods and, in particular, the methodology of computer engineering. This requires (basic) knowledge of special tools, as well as basic knowledge of neurology and psychoanalysis. After all, their models are to be used. It is always the first step in a computer simulation:[1] The elementary knowledge of the processes must be explained and described to the computer engineer by the respective experts of the processes. These must be well understood. The first development phase is therefore costly, but it cannot be avoided, because both sides (the computer technology specialists and the experts of the objects and processes to be simulated) do not know exactly what they need from the other side at the beginning. In this case, the neurologists and psychoanalysts do not know what is important and what is less important about their model in the first step of the simulation. The engineers are faced with a maze of information that is new to them, and to make matters worse, both sides use different jargons. Fortunately, I can abbreviate this difficult approach because we have been developing the project since 1999. We have spent many years working out what is important at the beginning of a simulation of the Ψ-organ in neurology and psychoanalysis. Therefore, I would like to describe the neurological and psychoanalytical knowledge for the simulation from my present point of view, i.e. from the point of view of an engineer. I allow myself to do this only under the supervision of my wife

---

[1] Neurons of the Ψ-organ (brain, neurons of the extremities, sensors) cannot yet be reproduced as synthetic tissue. Experiments with living neurons of a Ψ-organ are also difficult. So how can we experiment with them? The answer today is: by simulating them in the form of hardware and/or software. Dietmar Lange (1976) whom I refer to several times in the book, used both methods, because at that time it was still believed that only a technical reproduction, even a purely electronic one, is valid evidence. Today, it is known that the software-based proof is much more favorable, since it can be simulated in a wide range, up to the limits of physical possibility. Today, SiMA can only be simulated by software. The simulation imitates reality. There is no question that such simulations have limits. I will discuss this several times in the following. Additional note: the models that are simulated in natural scientific terms can also be emulated within certain limits. The model of the Ψ-organ could therefore also be integrated into robots.

D. Dietrich, V. Hartmann Cardelle, *Simulating the Mind II*,
https://doi.org/10.1007/978-3-031-69530-8_3

Dorothee Dietrich, who herself has had many years of psychoanalytic training, and my friends Klaus Doblhammer (psychoanalyst, doctor of philosophy and interdisciplinary scientist; see also the authorship of this book) and Volker Hartmann Cardelle (psychoanalyst, psychologist and PhD student of philosophy; see also the authorship of this book). Many might say that such an explanation has nothing to do with a psychoanalytic explanation, because the model of the Ψ-organ we have developed is only a technical model. *Precisely this is wrong.* Because I only have applied the methods of information theory that I learned and taught in my work as an engineer and professor of computer technology. The models we have developed are still the psychoanalytic models of the human psychic apparatus as they are presented in Sigmund Freud's theory of psychoanalysis or metapsychology. They are merely described (i.e. articulated) with different methods than those used in psychoanalysis. I have taken great care not to distort the theory of psychoanalysis, because otherwise the simulation experiments would lead to nonsensical results. Therefore, the following description should be seen as a different perspective and form of expression than that usually used by psychoanalysts.

In contrast to psychoanalysts, neurologists see themselves as pure natural scientists. Since, regarding information processing neurology is based on the laws of electrical engineering, the models of neurology can be applied directly in the Ψ-organ. Thus, the elaboration of neurological knowledge relevant to the Ψ-organ, in contrast to psychoanalysis, does not pose any particular difficulties in the following subchapter.

But before I go into the neurological knowledge that is relevant for us, as a computer scientist I have to correct a misunderstanding that I have often found in publications, but even more often heard at conferences. I will explain it using the example of a publication by Marianne Leuzinger-Bohleber. Freely quoting from Leuzinger-Bohleber (2008, p. 39):

> In the interdisciplinary colloquium mentioned above, we had to painfully experience again and again that we often do not speak the same language, that we use different concepts despite analogous terms, and that we feel bound to divergent traditions in the theory and philosophy of science. We need a lot of persistence, a lot of tolerance, to really enter into conversation and thereby question our previous ways of thinking, which is a prerequisite to really move forward to a deepening of disciplinary knowledge.

With this comment she hits the nail on the head. Precisely for this reason, this chapter is fundamental if one wants to understand how the *bridge* between the psychic apparatus and the nervous system can be worked out in a natural scientific way and how the Ψ-organ can be simulated. However, a few sentences later she falls into the very trap she has pointed out (again freely quoting):

> The brain is an organ, but it is not isolated; it is connected in many ways to the other organs of our body. These connections are all too often overlooked, especially by people who like to imagine the brain and its functioning like a computer.

Marianne Leuzinger-Bohleber is apparently not an expert in the field of computer technology and therefore certainly does not know the definitions[2] of the term *computer* used by computer engineers. It is therefore understandable that she not only uses this term in a colloquial way—which is fraught with ambiguities—and thus cannot but draw false conclusions from it (Dietrich et al. 2012, p. 123). She does not scrutinize her use of the term *computer*. She equates the *abstract term computer* with a technically realized computer, i.e. a thing made of sheet metal, plastic, and silicon, such as a desktop PC, laptop, or iPhone. But this is not correct. A comparison between abstract psychoanalytic functions and realized objects made of sheet metal and silicon is like comparing apples and oranges. A computer in the abstract sense can be described, according to the formulation of *information theory,* as an (abstract) entity, function, etc. that receives, processes, stores, and transmits information (Wendt 2013). Thus, by definition, there are both *artificial* and *natural computers. The brain* that Marianne Leuzinger-Bohleber talks about (i.e. the Ψ-organ in my terminology) *is a natural computer.* How a computer performs its task, what hardware it consists of, is not part of the definition of a computer. The computer engineer knows that, from the point of view of the natural sciences, a computer always addresses two "worlds", that of *hardware* (neurology, electronics, etc.) and that of *information theory.* These two worlds are closely linked, but this does not mean that they are based on the same scientific laws and description methods. On the contrary, both "worlds" require different points of view and thus different formalisms. But both "worlds" must be connected in a formalized way. Statistical methods, as applied by Mark Solms in Solms and Friston (2018), are only useful if their interoperability is guaranteed.[3] In the following, it is important for me to convey an understanding of these two "worlds" and how they can be connected on the basis of natural scientific thinking.

3.1 Neurology Plus

When people claim in discussions that the psychic apparatus cannot be compared to computer software and that the brain is not a computer, similar arguments appear again and again. Just consider the question of what the psychic apparatus has to do with software. A simple explanation is not possible, but I can give a clear and understandable answer. The basic problem is that there are many logical relations,

---

[2]I deliberately speak of definitions in the plural. As long as I was working in the international standardization commissions like IEC, IEEE, CLC, CEN, VDE, OVE, etc., I had to deal with the different definitions again and again. For financial reasons, we never achieved a unified formulation. The (axiomatic) formulations could only be achieved for specific technical disciplines. I worked mainly in the field of automation and therefore know its definitions best.

[3]In Solms and Friston (2018) the task is mainly seen in pointing out possible correlations, which then have to be verified experimentally. Yet no such an experimental verification is provided, which is why I cannot accept this elaboration in this way.

properties, models, and areas involved. Therefore, I cannot treat this topic in a closed context, but have to address it several times, depending on the progress of my explanations. The additional problem is that the confusion seems to be burned into the minds of many people. Moreover, I have the feeling that this is especially true of scholars and scientists who are particularly rooted in one and only one of these areas and therefore use the terms of the other areas too carelessly. Therefore, I would like to address a crucial point here and show on a scientific basis that it is irrelevant what hardware is used as the basis for an information system. Why is this the case? *Information symbols can be generated, transported, processed or calculated analogously, digitally[4] (in this case binary[5] or ternary[6]) or, as in the case of humans, on the basis of the threshold logic of the neurons.* This theorem of information theory is both fundamental and far-reaching. Its consequence is that from the point of view of symbolic information processing it is irrelevant whether I have an information system like an artificial computer, which is based on a binary electronic circuit, or a brain, which is based on neurons and is therefore a threshold logic—as long as the hardware is sufficiently capable. A consequence of this is that any information system (therefore also a Ψ-organ) can be simulated, i.e. its behavior can be reproduced and tested on an *artificial computer*. To understand this, it is necessary to have some understanding of the threshold logic of neurons and to compare it with binary electronics. What are neurons and how do they differ from electronic circuits? How do they work in terms of information processing? After all, that is their purpose.

It goes without saying that neurology is an extremely broad and exciting field, especially when clinical aspects are included. Here I have to limit myself to a narrow part of neurology: the information processing in neurons. Why, will hopefully become clear below. I want to move on a high (i.e. as simple as possible) level of abstraction. Many details of neurology are not necessary for the understanding of what I am aiming at in this book. Also, for the mathematical explanations (which go far beyond neurology, and which is why I have chosen the title *Neurology plus*) within information theory, which is essential for computer technology, I would like to refer to the corresponding publications of our research group, and especially to the nineteen dissertations and many diploma theses that have been written so far in the context of the SiMA project. Biochemical, physiological, membrane-specific considerations, the knowledge about the differences between different neuron types must be ignored here. Only the informational aspects are relevant, i.e. those that are

---

[4] Compared to analog technology, digital technology is based on discrete time values and discrete variables of various types, while analog technology works with continuous time values and variables.

[5] The binary system, also known as the dual system, is a number system in mathematics based on two digits.

[6] The ternary system, also known as the triadic system, is a mathematical number system based on three digits, not two as in the binary system.

directly relevant to the simulation of the Ψ-organ.[7] It is of interest, for example, that neurons generate electrical signals, what the circumstances are under which a signal is generated, how neurons transport information, how they store it, and how new information is extracted from old information. The physical, biochemical, electro-chemical, and physiological principles underlying neurons are beyond the scope of this discussion.

The study of networks of neurons—in the following I will only use the term *neuronal networks*—requires the inclusion of their temporal and spatial components. However, for the time being this is beyond the scope of SiMA, so I would like to refer the interested reader to the excellent book by Hawkins (2004, p. 23). In the following, it is only important for me to establish the connection between neurons and information.

The receptors (sensors) of the human being have the task of converting a physical quantity, such as light intensity or an accumulation (concentration) of substances such as gases or hormones, into an electrical signal and making these signals available at their outputs. These electrical signals are the carriers of information. Actuators (muscles and glands) have the opposite task of receptors. They process the signals from the neurons and convert them back into physical and chemical quan-tities. Neurons are phylogenetically related to receptors (Kandel 2006, pp. 233, 265). Therefore, both share common structures and exist solely to capture information at their inputs, manipulate it, transmit it, and optionally store it. From the point of view of engineering, I therefore count all *neurons* of the body including its *receptors* as part of the human *nervous system*.

These considerations are so fundamental to me that I would like to summarize them in a rigorous formulation as follows:

- The function of a receptor (sensor) is:

  - The conversion of a physical quantity, such as pressure, light intensity, etc., into an electrical quantity (electrical impulses),
  - the processing of information, and
  - the transmission of information.

- The function of a neuron is:

  - The processing of information,
  - the transmission of information and
  - the eventual storage of information.

- The function of an actuator (a gland, a muscle) is:

---

[7]There is no doubt that more profound theories are becoming increasingly important in this regard. I have spent many years studying such topics, but I have found that they are not crucial to the issue of bridging the gap between the nervous system and the psychic apparatus.

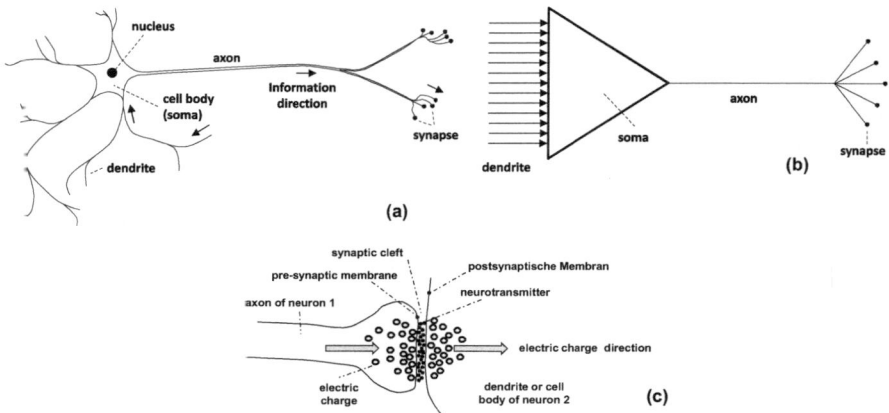

Fig. 3.1 Typical structure of a neuron. (**a**) Schematic (Meyer-Waarden 1975, p. 3); (**b**) informational representation (Lange 1976, p. 3); (**c**) synaptic transition between neurons

- the processing of Information and
- transformation of electrical quantities into other physical quantities such as force, production of biochemical substances such as hormones, etc.

At this point I would like to interject a thought. At a workshop, a scientist and biologist protested against my statement that ants and flies have a lower intelligence than more highly developed animals by claiming that they are highly intelligent. Just think of their social structures and the behavior that results from them, or bees executing certain dances, following the line of a figure eight,[8] and so on. I tried to explain that this was exactly the kind of discussion I consider to be pointless. His understanding of intelligence was simply different from mine. In such a case it is a logical consequence that discrepancies arise. The intelligence that lies in the social structure, in the behavior of a swarm of individuals, is to be distinguished from the intelligence that can be observed due to active information processing in a neuronal system, a brain, an organ, or a computer, such as cognitive performance.[9] In the SiMA project, only *active intelligence* is considered, i.e. the performance of the Ψ-organ and the computer. How does the information processing take place in all the neurons of a human being or in the electronic circuits respectively? As I said at the beginning, I would like to describe this process from the point of view of a computer engineer.

A neuron (Fig. 3.1a, b) consists of the soma (the central cell body) with a nucleus, the dendrites, and the axon. At the end of the axon are synapses, which connect to different parts of the following neuron (Fig. 3.2). The synapses introduce electrical

---

[8] I will come back to the figure of eight in connection with Braitenberg in Chap. 4.

[9] By structural intelligence I mean how sophisticated the structure of an "intelligent" unit is designed. By swarm intelligence I mean the abilities of "intelligent" units that act in a combined way. These two principles of intelligence are not subject to the definition of active intelligence.

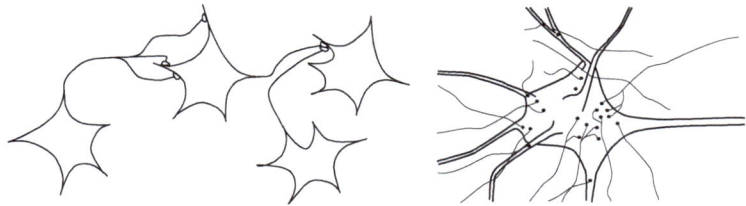

Fig. 3.2 Coupling of neurons. Left schematic representation, right possible synaptic couplings to a neuron

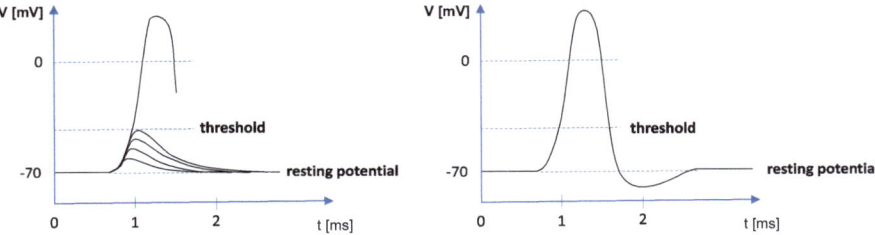

Fig. 3.3 Typical impulse of a neuron. On the left generation by the threshold crossing, on the right temporal course of an impulse

charges as input signals into the soma of the following neuron cell (Fig. 3.1c). The charges lead to the formation of electrical impulses in these subsequent neuron cells when they reach a certain electrical potential (threshold) (Fig. 3.3 left). This is called firing of a neuron. The result is an impulse and this impulse (Fig. 3.3, right) is conducted as a resultant output signal in the axon to its synapses. There, the signal is transmitted in the form of electrical charges to subsequent neurons or signal receptors such as muscles and glands. The structure, i.e. the *functional composition* of a neuron, is responsible for the complex information processing in the neuron. I will not go into these electrochemical processes in detail here and refer to textbooks such as Bear et al. (2016), Diener et al. (2019), Patestas and Gartner (2016).

Thus, neurons are threshold systems. This means that the charges introduced via synapses of preceding neurons or sensory cells into dendrites or directly into the neuron body lead to an impulse only if a certain threshold of electrical potential is reached in the soma (Fig. 3.3 left). This threshold depends on many factors and is highly variable (Eccles 1975; Küpfmüller and Jenik 1961; Lange 1976, p. 135). Also, when a new impulse can be generated again (after a refractory period[10]) depends on various conditions, such as the value of the resting potential, the type of neuron, and others. The maximum cut-off frequency of the impulses is about 1000 Hz, an extremely low value compared to today's electronic circuits.

---

[10]During the refractory period, the neuron cannot generate another impulse.

Synapses, the junctions between receptors, neurons, and actuators, can couple to different sites on another neuron or actuator: to the soma, to the dendrites, or even to another axon (this is called axo-axonic coupling) (Schmitz et al. 2001; Patestas and Gartner 2016). The flow of information in a synapse, from one neuron to another, is a complex process that can be influenced in many ways, including by hormones, which will be discussed later, because parallel to the information processing of the nervous system or the brain, information is also transmitted in the body via its hormones.

Information processing in neurons is not easy to understand, which is why I would like to explain it in more detail. The connection to the technical counterpart, the artificial computer, must also be worked out. For this purpose, basic knowledge of technical logic has to be introduced in order to be able to apply it to the description of neurons by means of information theory. In this way, the comparison between information processing in neurons and digital computer technology becomes practically possible and understandable. Ultimately, this leads to one of the decisive insights, namely that from the point of view of information theory, the hardware used does not play a role with regard to the psychic apparatus, as long as the hardware fulfills the requirements of the psychic apparatus. To put it concretely:

> From the point of view of information theory, it is irrelevant for the psychic apparatus whether the binary logic of today's artificial computers or the threshold logic of the neurons is available for an information system, as long as the hardware provides the required performance.

Why is that statement valid?

A distinction must be made between the hardware, i.e. the physiology of neurons and the hardware of artificial computers, on the one hand, and the respective applied mathematics that describes the information processing that takes place in them, on the other hand.

Computers are, by definition, systems that process information.[11] Before the term computer was coined, the German-speaking world spoke of *Rechner* (calculator) or *Steuerungseinheiten* (control unit). These were first mechanical, then electromechanical systems. Then vacuum tubes and finally transistors devices were built (Goldscheider and Zemanek 1971, p. 152ff, 194). For the future, many are waiting for the quantum computer, although it may be some time before it becomes a reality.

Analog and digital devices coexisted in the early days of electronic computers. With the use of transistors, digital computers became smaller and more powerful (in terms of information processing, especially in terms of accuracy (Steinbuch and

---

[11]The scientific term *computer* is thus to be understood as a purely abstract functional unit that manipulates, stores, and transmits data in one way or another. Strictly speaking, the term does not refer to the hardware. Thus seen, every Ψ-organ is a natural *computer*. In common parlance, the term "computer" means many things, the abstract function, but also the various "computers" that are designed as technical devices. Small wonder then that there is a lot of confusion. If we talk concretely about the devices of computer technology, for example the Apple computer, then these are in standardized formulation *computer systems*.

Rupprecht 1967, p. 328)), which helped to displace mechanical and analog calculators. Analog computers also did not achieve the same accuracy as digital computers.

The decimal system was generally used as the mathematical basis for both mechanical and analog computers. Digital computers (modern computers are almost always digital) are based on binary logic, which is technically the easiest to implement. Other forms of logic, such as ternary logic, have received no more than scientific attention (e.g. for easier derivation of certain algorithms) (Dietrich et al. 1990). For the disadvantage of binary logic, i.e. the necessity to convert the number systems into the easily readable decimal system as well as graphical symbols at the input and output end, useful solutions were quickly found.

It is important to work out the difference between the mathematical digital binary principle in technology and the threshold logic of the neurons in information processing. This makes it understandable why today's technical computers still use the simple binary logic and not the *much more powerful threshold system* of the natural computer (the Ψ-organ).

From the point of view of information theory, binary (digital) technology has only two states, which are defined as 0 and 1. Which voltage, which current, which frequency, which curve form etc. characterize (define) the two values 0 and 1 is independent from the technical realization. The mathematical laws are therefore easy to handle: *The information a and the information b result in the information c.* If we insert concrete numerical values of binary logic[12] for the variables a, b and c, we obtain, for example, if $a = \{1\}_2$ and $b = \{10\}_2$, the mathematical formalism for $c = a + a = \{1\}_2 + \{1\}_2 = \{10\}_2$ and for $c = a + b = \{1\}_2 + \{10\}_2 = \{11\}_2$. The mathematics, we know from our analog, ternary, or decimal world, can thus be easily translated into the world of binary logic and vice versa.

This is completely different in the world of threshold logic of neurons. There is no mathematical formalism that we, coming from our decimal or dual system, can easily apply to threshold logic. It is much more complicated to handle (Küpfmüller and Jenik 1961; Lange 1976). However, nature has created an extremely powerful computer based on this threshold principle, the Ψ-organ. We humans, on the other hand, still have to limit ourselves to the binary principle in technology because, unlike the threshold system of the neurons, it can be handled mathematically (i.e. calculated and implemented) without any problems. However, to repeat myself, from the point of view of information processing, it does not matter whether a natural or an artificial computer is based on binary logic or on threshold logic, which can be shown in the following after the explanations given so far and which will be theoretically considered and substantiated from another point of view in Chap. 5. *The different hardware principles concern only the realization, i.e. questions of size, energy consumption, whether it is a living system or not, etc., and they concern the method of mathematical description used.*

---

[12]The index 2 in $\{01\}_2$ indicates that two-value logic is used. $\{01\}_2$ is the two-value logic in which there are only the digits $\{0\}_2$ and $\{1\}_2$. Accordingly, in two-value logic $\{1\}_2 + \{1\}_2 = \{10\}_2$.

Fig. 3.4 Excitatory and inhibitory synaptic action on neurons. (**a**) Inhibition and attenuation on a neuron; (**b**) antagonistic inhibition; (**c**) antagonistic coupling of neurons in the eye for contrast enhancement

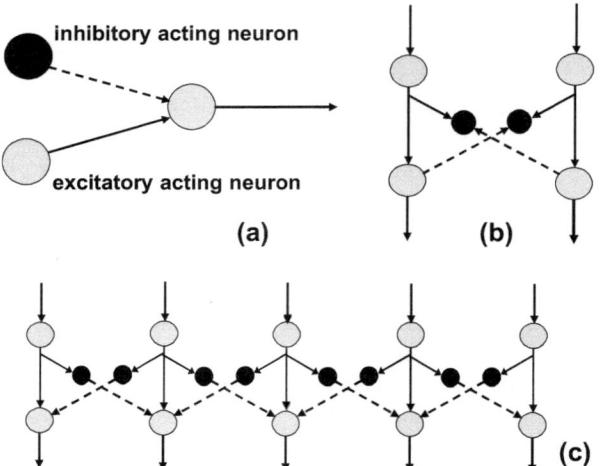

As shown in Fig. 3.3, neurons generate an impulse when synapses from sensors or other neurons bring in sufficient charge, i.e. when the threshold is exceeded. This is called excitatory coupling. Special synapses, on the other hand, can have an inhibitory effect. If a neuron has both excitatory and inhibitory couplings with other neurons, as shown in Fig. 3.4a, it depends on the sum of the charges of all acting synapses whether the threshold is crossed, and an impulse is generated.

Figure 3.4b shows a wiring that can be found, for example, in the retina of the eye (Lange 1976, p. 26; Jung and Kornhuber 2013, p. 49), a so-called antagonistic inhibition, on which the enhancement of the contrast of images relies. The mode of action can be developed three-dimensionally by calculation or experiment (Lange 1976, p. 63).[13] It can also be illustrated in a simplified way with 2D graphics as shown in Fig. 3.4c. If one assumes that the left and right neurons on the input side of the figure receive a weaker signal than the middle neuron, the inhibitory neurons on both edge sides have a stronger effect than that of the middle branch. The information of the middle branch has higher values compared to the branches at the edges. The eye (I do not only mean the purely optical system of the eyeball, but also the first neuron layers behind the photoreceptors) thus captures reality with greater contrast than it really is (Foerster 1993, p. 53).

As can be seen from the right curve in Fig. 3.3, the information transmitted by the neurons is not contained in the shape of the curves, their amplitudes, or the

---

[13] Based on the experiments of John C. Eccles (1975) and the model designs of Karl Küpfmüller and Jenik (1961), Dietmar Lange (1976) simulated six neurons in hardware and used them to simulate neuronal circuits based on the actual pulse code modulation in neurons. He thus emulated nature in a much more detailed (i.e. less abstract) way than is generally the case in the design of artificial neuronal networks. As a result, he arrives at results (as shown in Fig. 3.6) that are generally suppressed by the abstract models commonly used. However, these suppressed behaviors can lead to erroneous conclusions.

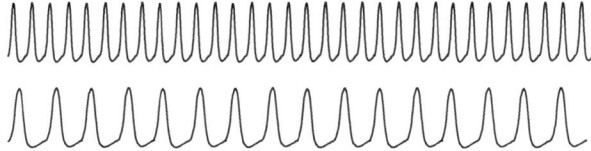

Fig. 3.5 Typical impulse sequence in neurons. High impulse rate at the top, lower impulse rate at the bottom

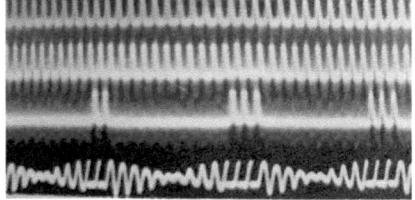

excitatory acting impulses

inhibitory acting impulses

burst-like output impulses

membrane voltage

Fig. 3.6 Typical impulse sequence in neurons according to Lange (1976, p. 152)

amplitudes of the currents flowing in the neurons, because they are always more or less the same for the corresponding neurons. From these parameters, a neuroscientist can only identify the type of neurons and, more importantly, their abnormalities and diseases. Neurons encode (package) their information solely via the time interval between impulses or via the density of the impulse sequence, which is called rate coding[14] (see Fig. 3.5). In electrical engineering, this is known as modulation of the electrical carrier (i.e. the electrical pulses) of the transmitted information.

For example, if a muscle receives a high pulse rate (Fig. 3.5, top), this will result in a strong contraction; a lower pulse rate (Fig. 3.5, bottom) will result in a correspondingly weaker contraction. Strong light incidence at a light receptor leads to a high pulse rate, weak light incidence to a lower pulse rate. Assuming a circuit similar to Fig. 3.4a, where an excitatory input is dampened by an inhibitory one, an output is produced as shown in Fig. 3.6. This was simulated by Lange (1976, see p. 152) based on the model of John C. Eccles (1975) and Küpfmüller and Jenik (1961). The mean rate of the output impulses is drastically reduced by the inhibitory input in the present experiment, which in this case even leads to burst-like output signals.[15]

Figure 3.6 also shows the membrane voltage of the neuron. The membrane voltage can be used to explain why these individual bursts occur, rather than a simple rate reduction, but this cannot be explained here. The reference to membrane voltage is only meant to demonstrate the complexity of information processing in neurons, which can also be technically simulated and studied in detail. Perhaps this will make it clearer what consequences a simplistic abstraction of a neuron can have.

---

[14] In general, it can also be referred to as Pulse Code Modulation (PCM).

[15] Bursts are short pulse sequences after longer pulse breaks.

Fig. 3.7 Forward excitation (Lange 1976, p. 161). Left: neuronal circuit; right: output frequency response as a function of input frequencies

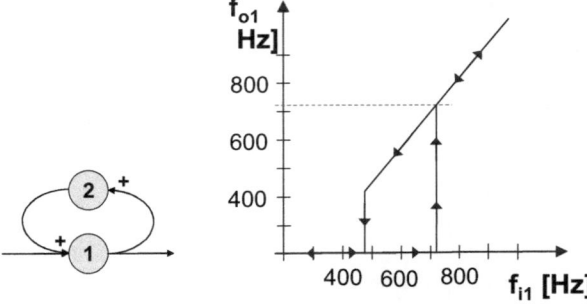

What makes neurons so enormously effective at processing information? To understand this better, two experiments from Dietmar Lange[16] (1976, p. 152) should be mentioned (Fig. 3.7).

Figure 3.7 on the left shows the positive feedback of two neurons. The technical term for this process is positive feedback via an intermediate neuron (interneuron). In Fig. 3.7 this interneuron 2, feeds back the output signal of neuron 1 and thus introduces additional charges into the soma of neuron 1. The behavior of this functional unit is shown in Fig. 3.7 on the right. The output frequencies f_{o1} of neuron 1 are plotted against the input frequency f_{i1} of the same neuron. Up to about 700 Hz, the output frequency f_{o1} remains at 200 Hz despite the increasing input frequency f_{i1}. Then, positive feedback begins, i.e. the output frequency jumps abruptly to a high value slightly above 700 Hz, which then continues to increase linearly as a function of the input frequency f_{i1}. When the input frequency f_{i1} is reduced again, the output frequency f_{o1} decreases linearly until the input frequency reaches a value of about 475 Hz. At this point, the output frequency drops abruptly back to 200 Hz. If in the present experiment an input frequency f_{i1} of about 600 Hz is assumed as the base frequency at which the output frequency f_{o1} is 200 Hz, then a brief increase in frequency of more than 100 Hz will result in an output frequency f_{o1} of more than 700 Hz. This output frequency then behaves linearly with respect to the input frequency until the input frequency f_{i1} falls below 450 Hz. The state behavior is to be seen as the storage of 1-bit of information[17] and reflects a model for the short-term storage of the Ψ-organ (for only a few seconds), which I will come to several times. First, however, I must draw attention to another experiment with a special neuronal circuit.

---

[16] Dietmar Lange uses the variable *frequency* instead of *data rate* for his experimental tests, which ultimately amounts to the same thing (Lange 1976, p. 64).

[17] 1-bit information is considered to be the smallest unit of information. From this it can be concluded that all possible information values can be represented on the basis of 0 and 1. Furthermore, it can be concluded from such considerations that the *AND*, *OR*, and *NOT* functions belong to the mathematically simplest and smallest information-processing functions in logical algebra. Other higher order information processing functions can be reduced to them.

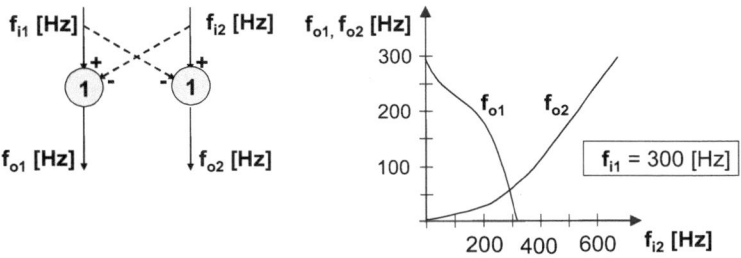

Fig. 3.8 Forward inhibition (Lange 1976, p. 157). Left: circuit; right: output frequency responses as a function of input frequencies

Consider the circuit shown in Fig. 3.8 and thus a simplified circuit of Fig. 3.4b: Two neurons inhibit each other via interneurons, a so-called mutual forward inhibition. The frequency dependencies are again highly nonlinear, as shown in Fig. 3.8 on the right. The output frequencies f_{o1} and f_{o2} are plotted against the input frequency f_{i2}. The variable f_{i1} is held constant during the experiment. If the input frequency f_{i2} is increased, the output frequency f_{o1} decreases and the output frequency f_{o2} increases. However, due to the pulse modulation, these dependencies are not linear throughout, but non-linear, and the shape of the curves changes as the input frequency f_{i1} is also varied.

The behavior of such *neuronal circuits*[18] (Figs. 3.4, 3.7, and 3.8) can therefore vary flexibly over a wide range. While the behavior of electronic, digital circuits can usually be easily inferred from their structure, this is by no means the case with neuronal circuits. Their structure is not as illustrative as that of electronic, digital circuits. You have to know the operating points of the individual neurons, take into account the influence of hormones on the synapses, and much more. Just consider that up to 1000 synapses are coupled to a single neuron (as shown in Fig. 3.2, right), that neurons can have a large number of dendrites, that neurons are often highly parallel, and that humans have an estimated 10^{10} neurons according to Antonio R. Damasio (1994, pp. 29–30) (according to Jeff Hawkins three times as many (Hawkins 2004, pp. 33–34)). You can perhaps imagine that such an entity not only possesses an enormous data processing capacity, but also that this cannot yet be adequately comprehended mathematically. In the near future, this will probably only be possible through simulations (Blue Brain Project 2005). The neuronal networks used in engineering and industry (see e.g. (Hawkins 2004, p. 23)) are still subject to

---

[18] I introduce the term *neuronal* circuit, which denotes an interconnection of neurons, because in artificial intelligence the term *neuronal network* is generally used for artificial circuits based on a strong abstraction of real neurons. For example, neuronal circuits does not use information transfer based on pulse code transmission, which does not account for certain behavioral phenomena such as bursts.

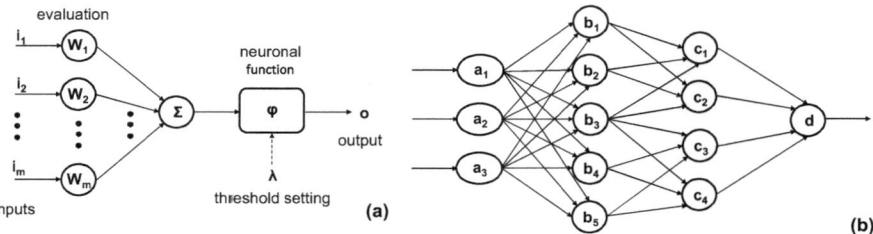

Fig. 3.9 Neuronal networks of artificial intelligence. (**a**) Model of a neuron abstracted to its essential communication functions; (**b**) neuronal network

strong abstractions[19] (Fig. 3.9)—i.e. simplifications—in order to keep them mathematically and technically manageable (e.g. via simulations).

In Fig. 3.9, the left diagram shows the representation of a neuron as an example of how the abstraction of a neuron can be made in the sense of artificial intelligence. The inputs i_1 to i_m can be analog values within certain bandwidths. While the following weights W_1 to W_m can stand for whether a synapse acts in an excitatory or inhibitory manner and how it is valuated with respect to the other inputs. The following sum character (Σ) reflects the type of electrical charge of the soma, and the result acts on a threshold function ϕ, which of course can be assumed to be arbitrarily linear or nonlinear, depending on the threshold setting. The output a of the abstracted neuron corresponds to the input.

The right figure in Fig. 3.9 shows a possible neuronal circuit, where each circle represents a neuron of the left figure. There are now a huge number of variants used in science and practice. Entire conferences are devoted to this topic. However, it must always be remembered that such abstractions reflect the performance of human brain neurons only to a limited extent. For example, I am not aware of the bursting behavior shown in Fig. 3.6 in any practical artificial neuronal network. But do such differences really matter, given the SiMA project's goal of bridging the gap between the neurologically described unit and the psychic apparatus? In order to answer this question, and finally to show that for the simulation of the Ψ-organ it does not matter what kind of hardware or what kind of logic (the digital or the threshold logic) is used, in the following the models of the threshold logic are compared to the binary logic. One could still use other mathematical principles like ternary logic, but this does not provide any additional benefit in terms of the goal of SiMA. Theory allows switching between binary and ternary. The differences lie in the possible efficient realization and the technology to be used, as well as its performance in terms of overall processing speed. Nature has chosen threshold logic. And for good reasons, but they can be better evaluated from a biological and physiological point of view. The simplest and cheapest logic to realize in a technical sense today is binary logic.

---

[19]Note—I repeat myself—that abstraction also means that the accuracy of a description is intentionally (often even dramatically) reduced.

This has not only technical but also mathematical reasons. A complete mathematical formalism for the threshold logic of highly complex neurons does not (yet) exist. In other words: One could, with a lot of effort, develop abstracted mathematical functions that come very close to the behavior of natural neurons. But then they would behave so extremely non-linear that one could not compute with them as easily as, for example, with binary or ternary logic. The only conceivable solution today is simulation. If the threshold logic is to be replaced by a binary, ternary or other logic, this can only be done by programmed simulations. One possibility would be to realize a neuron as a function in the form of a program and to multiply, to modify and to mesh it as often as necessary until one gets the neuron network one wants to study or apply. But in order to use this efficiently for simulating the Ψ-organ, tools would have to be developed, which could be done today, but the motivation and the money are lacking.

The above statement that binary and ternary logic as well as threshold logic are interchangeable from an engineering point of view for the simulation of the brain becomes more understandable by the following comparison of binary logic with threshold logic. In Chaps. 5 and 6, this statement will be justified again from a completely different perspective in the context of engineering. Some readers may need some patience to get a better understanding if they find the following too technical. Nevertheless, I see no other way to explain it. We have to dive a little bit into the world of electronics.

The principle of threshold logic of neurons has been illustrated above. Before contrasting it with the binary logic of electronics, it is necessary to discuss the basic concepts of digital electronics and explain the necessary terminology. This little excursion will take us into the world of binary (digital) logic, the basic principles of which are important for understanding the threshold logic of neurons from an information-technical point of view.

Figure 3.10a shows an electrical circuit in which a light (L) can be turned on and off by a switch (S). The switch (S) is open at rest. This representation is common in electrical engineering. The corresponding physical voltage curve V when the switch is closed is shown in the diagram in Fig. 3.10b. It shows that when the switch is turned on or off, there is a brief period during which the voltage oscillates around a

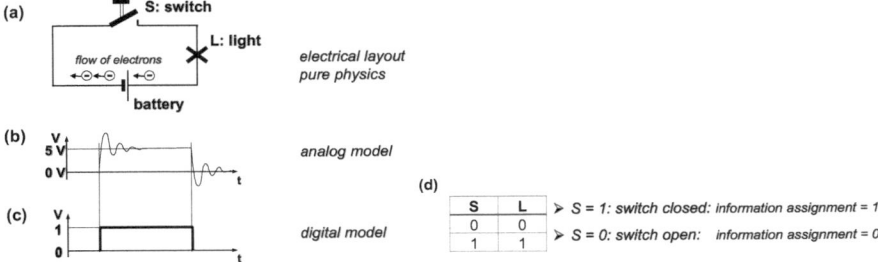

Fig. 3.10 Electrical circuit and associated voltage characteristics. (**a**) Electrical circuit diagram; (**b**) analog model; (**c**) digital model; (**d**) behavior table

certain value at which it eventually settles (in this case, 5 *V* or 0 *V*). In practice, however, it is also possible that the respective voltage value is adopted "creepingly", i.e. only slowly. This means that the temporary final state is not reached by oscillations around it. These different forms of transition from one voltage state to another depend on the physical characteristics of the switch, e.g. the type of switch, the type of light, the type of line, the switching frequency, etc., but this is only of interest for the realization of the switch.

In terms of engineering, the representation of Fig. 3.10b can be abstracted even further by assigning the information value 0 (mathematically formulated: $S = 0$) to the switched-off state and the information value 1 (mathematically formulated: $S = 1$) to the switched-on state. When the switch is closed ($S = 1$), the light is on, i.e. $L = 1$. When the switch is open ($S = 0$), the light is off ($L = 0$). The details of the transition processes are thus ignored. The representation of Fig. 3.10c is obtained. The information value assignment of the logical 1 (one) for the voltage value of 5 *V* and the logical value assignment 0 (zero) for the voltage value 0 V is arbitrarily taken. The assignment could also be done the other way round, i.e. the logical information value 1 (one) could be assigned to the voltage value 0 *V* and the logical information value 0 (zero) could be assigned to the voltage value 5 *V*. But these details are not of interest here, since they exclusively concern aspects of the realization as well as the technologies used. I will assume the first definition, which is shown in the table in Fig. 3.10d.

If the circuit of Fig. 3.10a is extended by an additional *switch* connected in series with the first *switch* (Fig. 3.11), the *light* will only shine when both switches S_1 and S_2 are on, i.e. both are closed and therefore both have the information value 1. This fact is shown by the behavior table in Fig. 3.11: $L = 1$ if $S_1 = 1$ and $S_2 = 1$, otherwise the right column (L) of the table in Fig. 3.11 is always 0. In computer technology, such a function is called a logical *AND* and is formally written as follows $L = S_1$ AND S_2. Inserting the values of S_1 and S_2 in a row into this equation gives the corresponding result for the value of L: the light L is turned on ($L = 1$) or off ($L = 0$).

By placing the second *switch* in parallel with the first, as shown in Fig. 3.12, only one switch (S_1 or S_2) needs to be switched *on* to make the *light* L shine. From the point of view of binary logic, we can therefore formulate $L = 1$ if S_1 or S_2 is equal to 1, or formally as an equation: $L = S_1$ OR S_2. The switch function is called a logical *OR*, and its behavior is shown in the table in Fig. 3.12.

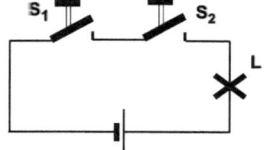

S_1	S_2	L
0	0	0
0	1	0
1	0	0
1	1	1

function: L = S₁ AND S₂

both switches S₁ and S₂ must be closed for light L to be on, i.e. L = 1

Fig. 3.11 With switches S_1 and S_2 evolved logical *AND* function, right beside it the behavior table

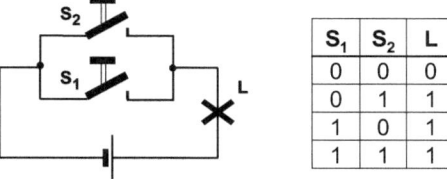

S_1	S_2	L
0	0	0
0	1	1
1	0	1
1	1	1

function: L = S_1 OR S_2

one of the two switches S_1 or S_2 must be closed for light L to be on, i.e. L = 1.

Fig. 3.12 With switches S_1 and S_2 evolved logical *OR* function, on the right-side corresponding behavior table

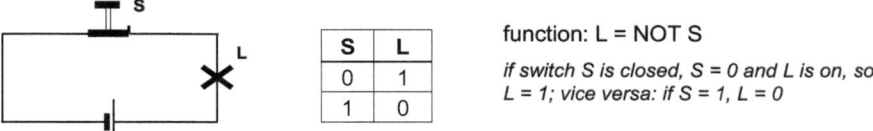

S	L
0	1
1	0

function: L = NOT S

if switch S is closed, S = 0 and L is on, so L = 1; vice versa: if S = 1, L = 0

Fig. 3.13 Logical *NOT* function

In order to be able to calculate all conceivable digital circuits, we need not only the two functions *AND* and *OR*, but also the function *NOT*, which is shown in Fig. 3.13. If we compare the switch *S* in Fig. 3.10 with the switch *S* in Fig. 3.13, we can see that the switch in Fig. 3.10 is open in the rest position, i.e. with $S = 0$ the lamp is not turned *on*. The switch in Fig. 3.13 works in the opposite way: In the rest position it allows the current to flow, which makes the light L light shine. So, for $S = 0$ you get $L = 1$. And only when the switch *S* is pressed, it is open and the light goes off and for $S = 1$ we get the assignment $L = 0$. The behavior of this negating *NOT*-function is shown in the table of Fig. 3.13. Formally we write $L = NOT\ S$.

In electronic circuits of computer technology, the slow mechanical switch is replaced by transistors that turn currents on and off. For example, Fig. 3.14 shows two kinds of *AND*-functions in (a) and (b), which can be drawn as a symbolic circuit as shown in Fig. 3.14c. In transistors, currents such as i_1 or i_2 can then be assigned the values 0 or 1. The corresponding table of Fig. 3.14d is therefore logically the same as that of Fig. 3.11.

The relationship between the different visualizations for the *AND* element, as shown in Fig. 3.14a–d, can be shown correspondingly for the *OR* and *NOT* functions. However, I do not want to give too many electronic details, but only show the corresponding symbolic functional representations, which will be needed later. Figures 3.12a and 3.14e show the logical *OR* and Figs. 3.13 and 3.14f show the logical *NOT* (without the internal transistor).

With these definitions of the three functions *AND*, *OR* and *NOT*, all conceivable digital, electronic circuits in computer technology can be mathematically described, calculated and developed on the basis of Boolean algebra, including memory principles. This is the reason why it can be used for comparison with other principles and why I mention it here. For this reason, neuronal circuits have been scientifically investigated to see if logical switches (in the sense of engineering) can be developed

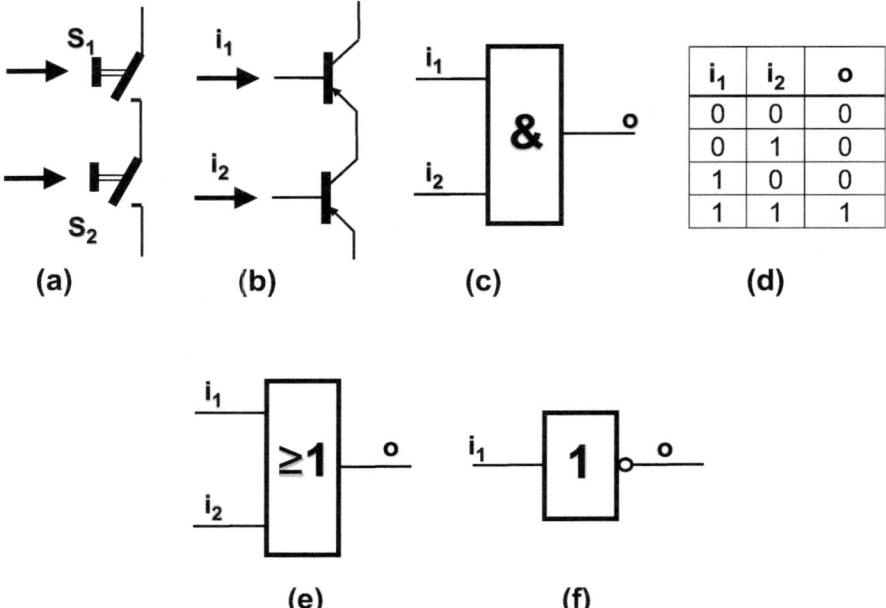

Fig. 3.14 (**a, b**) Transition from mechanical switch to transistor; (**c**) logic function representation; (**d**) Behavior table; (**e**) *OR* function; (**f**) *NOT* function. i_1, i_2: *input variables, o: output variable*

with them. Thus, the condition was created to transform the description method of digital logic based on the functional elements *AND*, *OR* and *NOT* into a neuronal one and vice versa.

> As explained at the beginning of this chapter, Dietmar Lange was able to evaluate, on the basis of the models of John C. Eccles (Eccles 1975) and Karl Küpfmüller (Küpfmüller and Jenik 1961), that, from the point of view of engineering, neurons are capable of generating many mathematical functions, including the three basic binary functions mentioned above and the memory function for short-term memory.
>
> The goal was not to develop electronic circuits with *AND*, *OR* and *NOT* functions on a neuronal basis. But to validate that, from the point of view of engineering, it is irrelevant whether an information system is based on a neuronal circuit principle or on classical electronic circuits with *AND*, *OR* and *NOT* functions. In other words, the question was whether the hardware of neuronal circuits is interchangeable with that of classical digital electronics.

The storage of 1-bit information for short-term memory is shown in Fig. 3.7. How can it be converted into a digital equivalent?

Figure 3.15 shows how the three basic functions *AND*, *OR* and *NOT* can be formed and described via neuronal connections (corresponding to Figs. 3.11, 3.12, and 3.13). Let me explain this in a little more detail.

The right diagram in Fig. 3.3 shows a neuronal impulse. Figure 3.15a shows the resting potential (*rp*) and the threshold (*th*) at which a neuron impulse (Spike) is

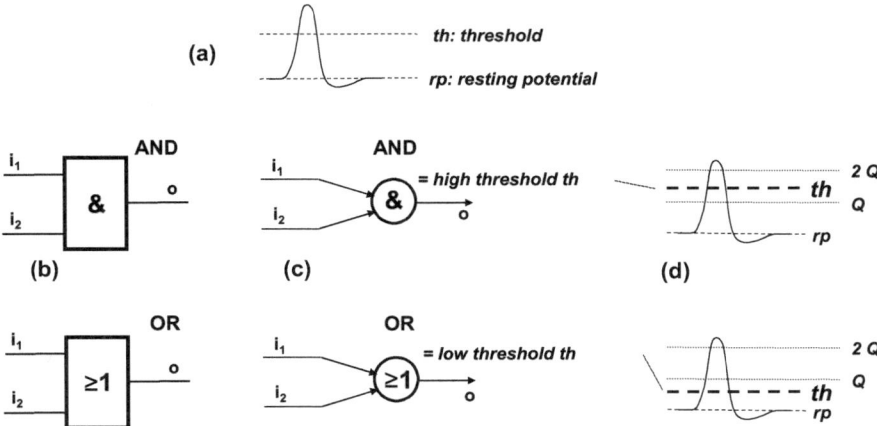

Fig. 3.15 Binary logic realized via threshold logic. (**a**) Definitions, (**b**) logical function representation, (**c**) neuronal circuit, (**d**) threshold representation. Q: charge in the cell body

triggered. This means that the synapses acting on the neuron must bring sufficient charges Q into the cell body so that the threshold is reached and the cell fires, i.e. generates an impulse.

Figure 3.15b shows the logical (symbolic) switch diagrams as shown in Fig. 3.14 for the *AND* and *OR* functions. The figures in column (c) of Fig. 3.15 show the corresponding neuronal equivalent. The thresholds in column (d) are fixed for the neuronal connections in column (c). If it is low (the lower illustrations of Fig. 3.15c, d, respectively), the charge of *one* input is sufficient for the neuron to fire. This results in an *OR* function. If the threshold is set at a high level (the higher illustrations of Fig. 3.15c, d), then both inputs must introduce charges at the same time for an impulse to be generated. This results in an *AND* function for the upper circuit.

However, one must consider that up to 1000 synapses (inputs) can be connected to neurons in the brain, and that the threshold setting can fluctuate, which was explained at the beginning of this chapter. This means that not only simple circuits like the one above can be realized with threshold logic, but also very complex ones, which can additionally be varied in time. Maybe this explains why no mathematical calculation methods could be developed for threshold logic like for the much simpler binary, ternary or decimal logic.

Figure 3.16 shows the operation of the memory circuit of Fig. 3.7 using both binary and threshold logic. To simplify the binary symbolic representation, an *OR* and a *NOT* function (Fig. 3.16a) are transformed into a *NOR* function (Fig. 3.16b). Two of these coupled *NOR* functions result in the representation of the 1-bit-storage function[20] of Fig. 3.16c. The corresponding storage capability of the threshold logic

---

[20] I cannot explain here how the circuit works. That would go too far into the technical details. I just want to show how memory units of electronics can be reduced to basic binary functions, which can be matched with the functions of threshold logic.

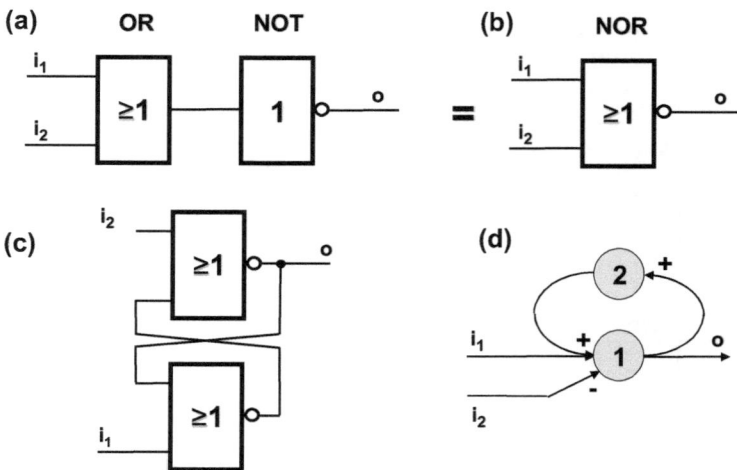

Fig. 3.16 1-bit memory circuit. (**a**) Realization of a *NOR* function based on an *OR* and a *NOT* function, (**b**) binary representation, (**c**) a possible 1-bit memory unit, (**d**) neuronal equivalent realized using two neurons

is shown in Fig. 3.16d. In contrast to Fig. 3.7, two inputs are specified here, with i_1 as the excitatory input through which the 1-bit information is stored. Input i_2, the inhibitory input, erases this information.

> The comparison shown between threshold logic and binary logic demonstrates that, from the point of view of engineering/information theory,[21] it is possible to switch back and forth between these two logical principles (one could add other principles such as ternary logic, the decimal system, etc.) without any loss of information.

From this it can be concluded that, from the point of view of information theory, it is irrelevant which kind of hardware is used, digital logic or the neuronal threshold logic.

The particular realization—i.e. which hardware and logic principles are used—only affects the effort in terms of the technology or physiological building blocks used, its efficiency, and other material factors. In other words, hardware (i.e. technology, physiology) and information theory must be seen as two different worlds that—as has been emphasized several times—must not be confused and mixed up. We must describe them differently and not confuse the methods of description. Even exemplary behavior from one of these two worlds cannot be transferred to the other. A transfer makes no sense, even if it seems reasonable. They are two different description methods, and there must be a well-defined interface[22] between them. I will discuss this in more detail later.

---

[21] From now on, the term information theory includes the term engineering.

[22] The term interface means that both sides of the interface, the side that is described in terms of hardware and the side that is described in terms of information engineering, cannot be defined

It becomes clear that the neuronal system processes information in a way that we can describe and thus model with binary logic. What humans have not been able to do so far, and in my opinion will not be able to do in the near future, is to develop a mathematical formalism for threshold logic in order to be able to develop the engineering of the Ψ-organ purely mathematically, just as one develops an artificial computer on the basis of digital (binary) logic. All this will be discussed in more detail in Chap. 5. There, however, I will choose a different approach. It has been used very successfully in computer engineering for the design phases in the last decades. Why not use this approach for the modeling of the Ψ-organ as well?

> The bottom-up method[23] used above to synthesize a system (such as memory units as in this case) is not the right one in computer technology to guarantee optimal solutions. This can only be achieved by the top-down approach.

Only the top-down method leads to an efficient development path. The bottom-up approach, i.e. starting from the building blocks (such as transistors, neurons, etc.), is acceptable only in exceptional cases to clarify certain questions.

What is the summary of the technical discursion?

> In the case of a model of an information system, regardless of whether it is an artificial computer that we have in front of us on the table or a biological system like our Ψ-organ, we can start from an informational description that is largely detached from the description of the hardware (physical/neurological). This means that for the concrete modeling of the psychic apparatus of the Ψ-organ we can free ourselves from the hardware, i.e. neurology, for the time being. We can imagine it as a roughly abstracted functional unit, but in such a way that the interfaces between the psychic apparatus and the neuronal system can be clearly defined, as dictated by the top-down method.[24]

3.2   Symbolic Information Processing of Patterns

The results of Dietmar Lange (1976), mentioned in the previous chapter, make it clear that the Ψ-organ is not a calculus machine like a modern, technically produced computer.[25] The receptors (sensors) of the different senses emit impulse series,

---

independently of each other and must be subject to an axiomatically defined connection/relationship.

[23] The terms bottom-up and top-down approach have to be used with some care, because they are sometimes used with different ideas (definitions). I use them from the point of view of computer engineering and computer science. Bottom-up means that a development starts with the smallest functions, which are always combined to larger functions like a modular system. In the top-down method, one starts with the definition of the largest function and breaks it down into smaller and smaller functions, decomposing it, so to speak, into smaller and smaller functions.

[24] Added by Volker Hartmann Cardelle: Consider the striking similarity between this sentence and Sigmund Freud's introduction of the psychic apparatus as a purely virtual entity (Freud 2017, p. 406).

[25] This wording is often misunderstood. This is why I have already addressed the topic several times and also mentioned Marianne Leuzinger-Bohleber (Sect. 2.3 and in more detail in the introduction

whereby the information, which is ultimately transmitted via the axon, does not lie in the form of the impulses, but in their temporal distance from each other—i.e. their respective momentary impulse frequency (this is why one also speaks of pulse code modulation). The shape of the impulse, on the other hand, tells the neuroscientist only to what extent abnormalities are present and should not be of interest here. But the temporal distance of the impulses is not the only piece of information. Neurons are spatially parallel. The spatial aspect of neurons is no less important (Hawkins 2004, pp. 56–57). When our eye sees something, about 6 million cones and about 120 million rods can simultaneously send the information in the form of impulses to the brain in parallel. I would like to call these simultaneously sent packets of information: *patterns*. Here we must realize that the eyes make sudden movements about three times per second, which is called a saccade. The eye fixates points and then jumps to the next (Hawkins 2004, pp. 56–57). The optical pattern, the image obtained by the physical eye, is thus composed of an unimaginably large number of temporally and spatially (parallel) resolved impulses that are fed to the Ψ-organ in the form of patterns, or more precisely, pattern packets. As Jeff Hawkins writes (Hawkins 2004, pp. 56–57), "Your conscious impression is of a stable world full of objects and people that can be easily tracked visually. But this impression is only made possible by your brain's ability to cope with a flood of retinal images (the pattern packets) in which no pattern is repeated. Natural vision, experienced as patterns moving into the brain, flows like a stream. Seeing is more like a melody than a painting." What does Jeff Hawkins mean by this? The flood of incoming information patterns (pattern packets), which are taken up in saccades, i.e. one after the other, must first be processed by the Ψ-organ in such a way that we can perceive a flowing event. The incoming patterns can be understood in terms of information theory, i.e. mathematically, as symbols, and this processing of information patterns can thus be described as a processing of symbols.

To understand this better, we need to be aware of an important aspect. We have to differentiate, on the one hand, how the Ψ-organ works, i.e. how it processes patterns, derives new patterns from them, stores them, passes them on, and, on the other hand, how we can describe all these processes symbolically (mathematically) and how we can understand in which functions of the Ψ-organ all this takes place. This is the only way to understand how this biological computer, our Ψ-organ, is able to perform such an enormous feat of information processing. It is the task of engineering to work out how to achieve this.

---

to Chap. 3), who is not correct with her formulation from the point of view of computer technology. She is right, however, insofar as the Ψ-organ in the sense of today's general understanding is not a computing machine like my or your laptop or like an iPad or any other corresponding device. However, this does not mean that the laws of physics, chemistry, or computer technology do not apply to the Ψ-organ (apart from the fact that engineering scientists have defined and even standardized the term *computer* as an abstract *function* that must be independent of any realizations, as I have already pointed out).

Fig. 3.17 My most favorite cup ("mon bol")

Every pattern is, in an abstract or mathematical sense, a symbol.[26] And whether the origin of these symbols is the eye, the sense of touch, the ears, the receptors (sensors) for pain or homeostasis, or a combination of different receptors, is irrelevant for the further processing of the symbols (Hawkins 2004, pp. 61–62). But it is important to understand that pattern formation is hierarchical, which ultimately means that patterns formed hierarchically close to receptors generally originate from a single receptor type, while patterns formed hierarchically far up (i.e. far from receptors) correspond to the combination of many sense organs (of different receptor types). The conscious impression I have of my favorite coffee cup (Fig. 3.17) does not only originate from the optical system, but also involves the patterns of touch as I grasp the cup, as my lips touch the clay, or the (patterned) sensation of the heat it radiates.[27]

"The brain knows about the world through a set of senses, which can only detect parts of the absolute world." (Hawkins 2004, p. 64). This means that the physical world must be defined as complex. Humans can only perceive a limited amount of its information. The rest must be inferred through abstraction. Thus, the brain abstracts the perceived objects via patterns, and we describe these patterns as symbols in order to describe and understand the process of perception. In terms of psychoanalysis (Solms and Turnbull 2002; Jacobson 1998, p. 58), the physical world is the external (complex) world. The world described by symbols is the inner world. It is reduced in information content, abstracted, and no longer complex. That is, all that the inner world knows about seeing, touching, or smelling are patterns, a finite number of

---

[26] The term symbol is not to be understood here in the psychoanalytic sense, but only in the purely mathematical sense.

[27] The subject of perception is a complex one. Oliver Sacks has studied it in a sophisticated way. That is why I would like to refer to one of his books in particular (Sacks 1985).

patterns that we describe in the model of the psychic apparatus through symbols (Hawkins 2004, pp. 56–57).

Strictly speaking, this way of looking at things is perhaps too abstract and simplistic, because if one examines the psychic apparatus a little more closely, which is the task of psychoanalysis, the question of what is to be considered complex and what is not complex requires a more careful analysis and thus a more differentiated view. The various functions of the psychic apparatus do not have access to all stored information at all times, but only to a very small part (the corresponding associations) at any given time. This is a crucial reason why also psychoanalysis distinguishes within the secondary process between preconscious and conscious content (see next chapter). Thus, the inner world does indeed need to be examined more closely in terms of its complexity. This is important for our model of the psychic apparatus, because it leads to the conclusion that we always have a complex world in front of us in our perception. This has serious consequences and leads to two further principles of perception, if one defines the acquisition of the pattern sequence as the first decisive principle of the perceptual process of the Ψ-organ.

One important point follows from what I explained so far. On the one hand, there is the sphere of neurology, which can be described in terms of physical, electrical, chemical, biophysical, and biochemical methods and laws. In it, electrical impulses are generated via receptors that transport information. Then there is the psychic apparatus, which can be described by the methods and laws of information theory. There, the pattern sequences are described in the form of symbols. And between these two areas there is a world, a realm, a dimension—whatever you want to call it—in which the symbols to be described mathematically are obtained from the electrical impulse sequences of the neurons (the world of neurology). In terms of description, this transition from electrical impulses to symbols can be seen as a kind of translation of languages. The psychic apparatus works with symbols. In accordance with the hierarchical structure, the reverse can also be understood in this way. The symbols (= pattern sequences) of the psychic apparatus must, on the descending path, ultimately be broken down again into electrical impulses for the individual muscle fibers and glands (desymbolized).

Symbolization, then, is the merging of information from different neurons into a single piece of information, perhaps in a single neuron, which has a temporal and a spatial dimension. The intervals between impulses represent the temporal aspect. But since the information generally comes from sensors arranged in parallel, and the channels for signaling continue to some extent in parallel, there is also a spatial dimension. Symbolization proceeds hierarchically in the Ψ-organ. The further this process proceeds from the receptor layer, the more detailed information is lost, the more abstract the symbolic information becomes, especially when receptors of different sensory organs are involved. The process of desymbolization works in the opposite direction, in parallel, to the process of symbolization.

To this topic, since it is rather difficult to understand, I will devote a separate subchapter in Chap. 6, which is mainly based on the scientific work of Rosemarie Velik (2008).

Symbolization and desymbolization thus occupy spaces between neurology (hardware) and the psychic apparatus (information-theoretical part). Where these spaces fit into the model of the Ψ-organ is mainly the subject of Chaps. 5 and 6.

But in order to understand why AI is still struggling to produce robots that can easily arbitrarily discriminate between different objects, and why we scientists in the SiMA project are still struggling in this regard, I need to elaborate on two other phenomena of the process of perception.

Random objects located in a room, and especially in an unspecified environment such as an open terrain, can be easily identified by humans. But appearance is deceiving, especially here! Humans can only perceive what they have seen before in a more or less "similar way". Humans have to "learn" their environment, whatever that may mean at first. The perceptual power of the Ψ-organ is therefore not limited to forming—or rather: working out—a smoothly flowing stream from the raw material of saccadically produced patterns, which, in mathematical terms, I call symbolic computation. Jeff Hawkins has already in 2004 (Hawkins 2004, pp. 75–76) described very well two further crucial abilities of the Ψ -organ, which are also essential for the performance of perception. One he calls invariant representation the other auto-associative retrieval. These are two phenomena that have long been known in psychology and pedagogy, but could not be technically reproduced at the time when Jeff Hawkins wrote his book. In other, more scientific, words: There were no models of the functions that produce such processes. One searched the literature in vain for useful solutions. The invariant representation was only taken up and solved by Martin Pongratz in the sense of the SiMA project for tasks in automation (Pongratz 2016). However, he was not familiar with the work of Jeff Hawkins. Martin Pongratz approached the topic from a different perspective and thought almost exclusively in the conceptual world (axiomatics) of automation. However, I find Jeff Hawkins' way of describing the problem so apt that I would like to adopt his way of presenting it in the following, even though he only pointed out the problem but, in contrast to Martin Pongratz, could not yet offer any experimental solutions.

How did Martin Pongratz arrive at his solution? The packaging industry called for the development of robots that could throw objects between production stations instead of moving them on fixed conveyors (e.g. for packaging cookies, pharmaceuticals, or similar goods). This would potentially save an enormous amount of money and make the logistics in production halls much more adaptable. A first scientific work in this direction was developed by Dennis F. Barteit (Frank et al. 2008; Barteit 2010). His robot had to catch a ball thrown from a distance of about 3–5 m. It was based on all classical principles, all physical and mathematical laws and algorithms known from the technology of flying objects were considered: aerodynamic forces, the spin effect, the surface properties of the flying object, and so on. The results obtained by Dennis F. Barteit were considered insufficient by the industry, although the work was scientifically excellent. I could understand the industry. The percentage of objects caught was poor, considering how many parcels the robot would have dropped in an industrial hall if the principle had been applied in practice. This was despite the fact that most of the robot's functions had been intensively optimized (Barteit 2010, p. 40), much better cameras had been integrated, and other refinements had been made (see for example (Akhter 2011)). In light of these results, the fact that a tennis player (Martin Pongratz was a professional tennis player) is able to adequately parry balls flying at more than 100 km/h became highly intriguing. What happens in a human being when they throw and catch objects?

We knew we had to move away from the classical approach of physical calculations. Based on our scientific results and all the scientific contributions worldwide, this approach simply could not be the right one. Humans rely on different principles when they throw and catch. They do not calculate the parabola of the throw, the spin effect of the ball, or take the surface roughness as a constant. They do not calculate the lateral flow of air, nor do they take into account as many nonlinearities as possible during the flight. The answer to how we humans function in this case is given by developmental psychologists such as Dornes (2001, pp. 87, 96). When a baby discovers the process of dropping objects, as every mother and father knows only too well, it tries to drop objects and shows great joy in doing so. This is an important phase of learning. The child develops a "feel" for moving objects. They learn and memorize the difference between slow- and fast-moving objects. But what exactly are they memorizing? Certainly not flying cups or spoons. But from the flying cup, the infant perceives moving edges, circles, colors, and so on. These patterns of motion create traces that are stored, though not the absolute locations, but the form in which the patterns move, which I will call trajectories[28] (Pratl 2006). Martin Pongratz (2016), assisted by Konstantin Mironov (2016), who launched many balls many times, recorded these motion patterns and programmed the trajectories into his robot. They were virtually imprinted on it as a memory. The robot's only task then was to determine an optimal trajectory from its memory, i.e. to find the trajectory among the many programmed trajectories that best matched the trajectory it had calculated itself. In this way, the robot knew where the ball would land. There was no need to know complicated laws of ballistics or aerodynamics. The robot simply used its "learned" knowledge as a basis. Although Martin Pongratz and Konstantin Mironov did not have time to optimize the technical details, the results of their first scientific design were already orders of magnitude better than the results of the classical approach of Dennis F. Barteit. It was astonishing, especially as we slowly understood which technical details we could have decisively improved.

To be honest, Martin Pongratz, Konstantin Mironov and I did not know how the Ψ-organ actually determines trajectories, nor how it recombines them. One has to consider that a computer can perform more than 1000,000 complicated computational steps in a hundredth of a second, while the Ψ-organ, roughly speaking, manages perhaps 50 or 100 neuron actions in succession in the same time. It is not enough to attribute everything to the parallelism of computing power, as Jeff

---

[28]From these considerations an important insight can be derived. The psychic apparatus processes information in the form of symbols; we call them patterns. Patterns can be: a static optical image, a sound, a weight, the movement of an object, a smell, the touch of a surface, and so on.

Hawkins describes in detail (Hawkins 2004, pp. 17–18). Much research is still needed here, but the work of Martin Pongratz and Konstantin Mironov shows crucial things: The Ψ-organ works with different approaches than today's common technology and shows better success. Technology should orient itself on what infant researchers have worked out.

In this sense, invariant representation means that humans do not store objects as patterns, pixel by pixel, regardless of whether they are derived from visual, auditory, olfactory, gustatory, or tactile patterns, because, extrapolated over the lifetime of a human being, that would be incomparably more pixels than the human being can possess synapses. The incoming pattern sequences are converted into internal, invariant representatives, which are again patterns, but characteristic, resting patterns. This principle has long been understood in the field of person recognition. It makes little sense to store fingerprints pixel by pixel, but only the characteristic shapes, such as branches of fingerprint lines, their relative distances, and so on. The same is true for passport photo software. It is no longer the pixel patterns that are interesting, but the relative distance of the eyes to the edges of the mouth, the relative size of the mouth, and many other characteristic features. This is the only way to recognize a face after many years or a face with or without a beard. When we see a person, we first see the stored, inner invariant representative of that person and determine (create) from it the image that does not contradict the inner one. And if it is an old friend whose face has changed a lot, we valuate this change negatively or positively, i.e. we experience a new learning process. Perhaps this helps to understand why different trustworthy people can perceive an accident differently and still insist that they are not lying and that they saw exactly what they say they saw. They must have seen the same event differently because their respective perception relies on different stored patterns.

But in this last example, a third phenomenon plays a crucial role: auto-associative pattern retrieval. Jeff Hawkins uses a very vivid example to explain this. We know our glasses. No matter how we see them, touch them, we immediately have our glasses in our mind's eye (Hawkins 2004, pp. 78–79). And this is what we do with all objects, just as I described above regarding invariant representation. When we see a package of nuts, it looks different from each side. Yet we recognize it as a package of nuts. In the classical approach, computers have stored images of them pixel by pixel. This means that every single image is different. This is not feasible if you want a robot to recognize a package of nuts on a table. You would have to store thousands of packages of nuts. And that would be the case for every single object that the robot sees at that moment. How can it do that computationally? Therefore, only a few features need to be used as a basis for pattern recognition. In the case of identifying a person, these are the distance between the eye lenses, the shape of the eyebrows, the position of the mouth, the size of the mouth in relation to the size of the nose, and so on. Obviously, the more features you define, the better the accuracy. Now here is the interesting thing: How is it that I can immediately recognize an old friend whom I have not seen for 40 years, despite his changes, while I cannot recognize other people at all? This is easier to understand when we enter the subject of psychoanalysis, which is the subject of the next chapter. This is the reason why in SiMA, as we will see later on, a strict distinction is made between perception and recognition. More on this in Chap. 6.

One last detail to conclude this chapter. How does the Ψ-organ work out characteristic sizes, if not on the basis of pixels? Of course, the cones and rods of the eye initially see only pixels, but even the first layers of neurons recognize edges, circles, triangular shapes, etc. in these pixel patterns (Hawkins 2004, pp. 45–46). The decomposition of an image detected by the receptors thus already takes place in the lowest neuron layers in the eye. Heinz v. Foerster was able to prove that these first neuron layers also intensify the contrast of the images, i.e. work out the edges better than they are initially perceived by the receptors in the eye (Foerster 1993, p. 45). The optical image is thus resolved and processed over many neuron layers, whereby characteristics are worked out. From this, the Ψ-organ—in a higher layer—calculates new patterns, five fingers become a hand, a hand and the lower joint become an arm, and so on, which in turn is stored in an abstract associative way and is then available for retrieval. This has the serious disadvantage that we do not remember the material reality exactly, i.e. in every detail, but we can from fragments of an image develop the whole picture. We work out our inner pictures (invariant representatives), so to speak, before we see them completely. It is not for nothing that different people see different things when they witness an accident. Everybody computes their pictures, which they associate from their memory with the immediate patterns. The advantages are obvious. First, a situation can be grasped extremely quickly. Few characteristics are enough to have an entire picture. Secondly, I need to see a package of nuts only once for a short time, and although I have never seen it exactly in the way it is lays before me, I do not have to "work out" a similarity, but simply associate the package of nuts from my inner stored images. And in this manner, I recognize my friend even after 40 years.

3.3   Psychoanalysis Plus?

There are two things I want to point out and keep in mind. First, I cannot explain the theory of psychoanalysis,[29] the so-called *metapsychology*, in general. I can only deal with those aspects that are crucial for the scientific project SiMA. Second, it quickly became clear in the SiMA project that we can only see psychoanalysis in the context of neurology and brain research in general. We cannot ignore the *interfaces* with other scientific fields. This is what I mean by the word *plus* in my title. That is why the following descriptions sometimes go a bit beyond the theory of psychoanalysis.

---

[29] In the following, I will speak only of psychoanalysis, for the sake of simplicity. It is also clear from this statement that the term psychoanalysis in the strictly scientific sense actually has to be differentiated more precisely. Sigmund Freud distinguishes three meanings of psychoanalysis: (1) the method of investigating unconscious psychological contents and processes, (2) the psychotherapeutic method, and (3) the theories derived from these methods, e.g. metapsychology (Freud 2020e, p. 203).

However, the *interoperability*[30] of the different fields is of utmost importance. From a scientific point of view, contradictions are not acceptable.

Second, I describe psychoanalysis—since I am not a psychoanalyst[31]—from the point of view of an engineer. I see the theory of psychoanalysis in this sense as a natural science that deals with the information functions of the Ψ-organ, its *information content* and the resulting behavior and experiences of a human being, and thus try to understand the human being as a process that is controlled by *physiological* and *psychological functions* of the Ψ-organ. *The computer engineer (less the computer scientist) learns in theory and practice to describe, analyze and develop mixed processes of hardware and engineering.* What I have learned and taught in this sense in my professional life, I apply in the following.

I have little knowledge of clinical methods of psychoanalysis. But they are not relevant for the development of a tool for the simulation of the Ψ-organ. It is up to other researchers who have the necessary expertise to decide how this tool will be used clinically. For this reason, the clinical field of psychoanalysis is generally excluded from SiMA. Needless to say, I have no doubt that in the future psychoanalysts will use the sophisticated tools for simulating the Ψ-organ for clinical purposes. But we are not there yet.

I will also leave aside the contradictory theories of the various schools of psychoanalysis.[32] I see these as detailed knowledge that is not necessary for a basic simulation. However, in the future these theories will have to be tested with the simulation tool to see if they are coherent or if they lead to contradictions that need to be resolved.[33]

There is another important point to make. Psychoanalysis has evolved since Sigmund Freud. The neurosciences in general have developed tried and tested scientific pillars that offer explanations that psychoanalysis is not in a position to provide. The simulation in the SiMA project sometimes requires differentiations that may be unimportant for psychoanalysis today. All of these considerations are included below, but always under the premise that they do not contradict classical psychoanalytic theory.

---

[30] *Interoperability* became an important standardized term in automation when it was understood that in computerized networks, standards alone are no longer sufficient for one device to truly "understand" another (Kabitzsch et al. 2002).

[31] However, I must emphasize again that my text is always proofread by at least three people, by Dorothee Dietrich (my wife and group psychoanalyst), by Volker Hartmann Cardelle (psychoanalyst) and by Klaus Doblhammer (psychoanalyst). All three are very critical in their evaluations to ensure consistency with psychoanalysis. In this way, I believe I am fulfilling the requirement for control by experts in fields in which I am not sufficiently proficient.

[32] These include for example the *Lacanian school*, the *Kleinian school*, the *schools of ego psychology*, of *object relations theory* etc.

[33] In a scientific sense, the results of the simulation experiments will, over time, lead to a unified, generally accepted model for all schools. However, I am deliberately not referring to clinical methods here.

In 1999, during a week-long boat trip on the River Lot in southern France, I talked with my brother-in-law, a professor for special education, about psychoanalysis. At some point I became ardently interested because I came to understand that already in 1923 Sigmund Freud in his *structural model* (Freud 2020e)—also called the *second topical model*—had developed the psychic apparatus as a *functional information unit*, which he divided into three main functions, the *id*, the *superego* and the *ego* that by their interoperation bring about psychic activity, e.g. thinking and planning of actions, and thereby ultimately human behavior. He thus had used concepts of information processing that computer technology would not define until decades later. Since then I have found it colossally impressive: Sigmund Freud developed from his analyses of mental phenomena and treatments of patients *functional information units* that control the physiological body. He distinguished between the *functions of the physiological organ* and the *functions of the mental system*. And Sigmund Freud really had nothing to do with automation, technology, and engineering. Norbert Wiener, Konrad Zuse, Claude E. Shannon, Heinz Zemanek all came up with similar considerations and definitions many years later. I was thrilled by this insight, which was fundamental for me, and in the following years I set out to investigate all 17 psychological and psychoanalytical schools that existed in Vienna at that time. *Which of these schools use methods similar to computer technology? Which schools have a similar approach to the development of functional models?* It was clear: *Sigmund Freud's psychoanalysis* (i.e. his metapsychology) was *the only one*. No other school had or has a model that is as functionally well elaborated as the *structural model*, not even remotely. The others had exclusively *behavioral models* models, behavioral (in the sense of the definitions of computer technology), including other psychoanalytic schools, although they do not call them that. But for the computer technician, it is crucial to first work out from behavioral models a functional model with main functions, sub-functions, sub-sub-functions, and so on.[34] *The functions produce the specific behavior that is desired.* This was the central theme of my professorships in Germany and Vienna, as well as my various visiting professorships and stays in Canada, South Africa, Russia, and China. And this is exactly what I am trying to work out here for the Ψ-organ.

Let me illustrate this with a simple example. The engineer thinks about a radio in a relatively simple way: For him, the radio is a *functional unit* with the *sub-functions* antenna, input amplifier, filter, and so on. These functions ultimately determine the *behavior* of the radio. For development, these sub-functions must be broken down into further sub-functions, etc., until small functional units—e.g. transistors, resistors, coils, etc.—are obtained, which can be realized and assembled. It is important to understand why certain functions are created in hardware or in software, i.e. as purely physical electrical components or in the form of information objects (i.e. *information function units*). The information objects are then part of a software

---

[34]The differentiation of functions into sub-functions and sub-sub-functions was not possible for Sigmund Freud, which he was aware of Freud (2021b, p. 370). This requires tools that he did not have. Today, simulation tools in particular make this kind of work much easier.

program of the microcomputer to be integrated. In this way, for example, certain filters in a radio can either be realized as hardware or—today much more cost-effectively and above all in better quality—as a *software function*.

Why should this method of structuring not work in the same way in psychoanalysis with the three functions developed by Sigmund Freud? It is undoubtedly more difficult because, first of all, a model of the Ψ-organ is very complicated. Secondly, in the development of radios, humans dictate the structure. In the case of the Ψ-organ, there is the object, the Ψ-organ, and humans must learn to describe it as a functional model. And third, if the model of a radio is wrong, the simulations and experiments will quickly reveal this, and it is then only a matter of diligence to find the errors. In the case of humans, the simulations and experiments are more extensive. And fourth, and perhaps one of the most important criteria, to get a feeling (knowledge) for modeling and simulating radios, one could start with simple, less complicated radios. The Ψ-organ with all its complexity has to be taken as it is.

Nevertheless, the simulation of the Ψ-organ must bring it to light: *Either Sigmund Freud was right, or the simulated system of smallest functional units resulting from his theories produces nonsense in the simulation.* The detailed structure will be the subject of Chap. 6 of this book. For now, however, let us return to psychoanalysis.

Computer engineers think primarily in terms of images and processes (information processes), whatever those processes may be. I will present the following in this sense. This kind of representation is more compressed and efficient than a textual one, and it is axiomatically unambiguous. However, one is forced to think into the graphics.

How can we understand the three functions id, superego and ego? Let us start with Fig. 3.18. Sigmund Freud spoke of an *external world* and an *inner world* (Solms and Turnbull 2002, p. 18). The external world includes everything physiological about a person and the *surrounding world*, or in scientific terms: everything material. The physiological can be divided into the neuronal system—including the receptors and actuators (to control muscles and glands)—which processes the information, and all the other organs down to the muscles and the skeleton. The inner world is the *psychic apparatus*. Sigmund Freud tried to develop a model for the connection between these two worlds, but he failed. He realized that both models

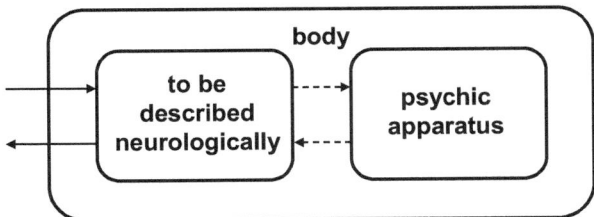

Fig. 3.18 Abstracted information system of the human being (information process), as we have to divide it into two blocks for the psychoanalytical and physical-chemical (neurological) description according to the theory of Georg H. Mealy (Chap. 5). Dashed line: defined in psychoanalysis as unknown transition

had to merge into one another (Freud 2020f), but he did not know exactly how this could happen. He wrote towards the end of his life (Freud 2021b, p. 428) more or less that: ". . . a direct relationship between both endpoints of our knowledge [i.e. both worlds] is unknown."

So, let us look at this problem of the bridge, or the interface, as we computer engineering scientists call it. And this is what I am concerned about. We are now in a position to solve this problem. It is important to realize that Sigmund Freud, despite the difficulty of formulating it precisely, understood even then what computer scientists understood much later: *The external world is governed by the laws of physics, the inner world by the laws of information theory.* It was not until George H. Mealy (elaborated in more detail in Chap. 5) that computer engineers were able to grasp this theoretically. For example, humans have somatically determined *drives* (external world), which are processed as *drive representatives* in the psychic apparatus (internal world) and trigger *complex processes* in the internal and external world. A human being perceives homeostatic deviations, imbalances, hormone accumulations, etc. in their body (external world) via *receptors (sensors)* for various types of information and reacts to them. Is blood sugar, blood pressure, oxygen saturation all right? A human perceives images and sounds through the receptors of the eyes and ears. Their hand touches objects. All this information becomes representatives for the psychic apparatus (inner world)—so it is also information, but now abstracted—which we can describe in the form of *symbols*.[35] The psychic apparatus can work with these symbolic quantities of pain, thirst, visual images, etc. However, this also makes it clear that there are inaccuracies and falsifications in the recording and conversion of physical/chemical quantities into electrical information quantities. And not only with negative effects, but also in an advantageous way, as explained in the previous Sect. 3.1: For example, the special wiring of the neurons leads to an intensification of the contrast in the perception of the material reality. In summary, everything that is processed in the psychic apparatus is only information about the material world, which always corresponds more or less to material reality, *but can never represent it exactly.* The symbols (information) are abstract quantities. What we see, hear and feel is *not* the material reality, but always only an *abstracted* reality *recognized* via our sensory system. The process of *recognition* is very *complex* and

---

[35] At this point I must ask the reader to perform a difficult mental operation, which will be explained in detail in Chap. 5. A distinction must be made between what actually happens physically and chemically and how we are able to *describe* it. This is what the Mealy theory explains. From a physical and chemical point of view, everything is one process, but this process is so complex that we have to *describe* it as if it were split into a process that can be described by physical and chemical methods and laws, and a process that can be described by the methods and laws of information theory. Within our head no such transformation into symbols as in technology occurs. But it is only by means of such a *mathematical transformation* that we can describe and ultimately simulate and emulate the information processes that take place in our brain. And this is how Fig. 3.18 should be understood. The neurological and mental process are essentially one process which we have to treat, based on the theories of Sigmund Freud *and* George H. Mealy, as if it consisted of two processes that interact with each other. This is not only difficult to understand, but it also causes great mathematical problems for many people. If you read Chap. 5, it will become more understandable.

involves much more than the perception of sensory information. Today's robots and computer systems do not yet achieve human *recognition* in the sense of the SiMA definition, but only *perception*, which will be a central topic in the following.

I have not included the human hormonal system. Strictly speaking, it too has to be included in the study of information processing. However, in the first stage of development we have to neglect it in order not to push the complexity of the relationships too far, which is why it has been left out of SiMA so far. The reason why this is not a problem in the first approximation will be explained later.

But before we come to the three functions of the psychic apparatus mentioned above, which form the second topical (Freudian) model of the psychic apparatus (the structural model), I would like to mention, for didactic reasons, the first topical model, which Sigmund Freud developed long before the second topical model (Freud 2017, p. 407). Here Sigmund Freud divides the psychic apparatus into two areas, the area of the unconscious on the one hand and the area of the preconscious[36] and conscious on the other hand (Freud 2017, pp. 410–411). The unconscious was imagined to work according to the primary process and the realm of the preconscious and conscious according to the secondary process (see also Freud (2017, pp. 458ff, 465ff)).

The terms *the* unconscious, *the* preconscious and *the* conscious imply functions which are *unconscious, preconscious* or *conscious*. But the properties unconscious, preconscious and conscious cannot be assigned to functions from the point of view of the technical information theory. They are just *properties* of different pieces of information, which are processed and stored in the *functions* of the psychic apparatus. This way of defining the terms is compatible with the theories of computer technology and informatics.[37] Please note that this does not mean that we are changing the content of Sigmund Freud's theory. As I will explain in more detail later, the differentiation between the *functions* and the *properties of the information* (such as *conscious, preconscious* and *unconscious*) in terms of the information theory of computer technology leads to a *functional model* and a *behavioral model* (of the information of the psychic apparatus). Both models complement each other and neither model makes sense without the other. Thus, in this sense, in the psychic

---

[36]Preconscious is the information that is not conscious but can become conscious. A simple example: Imagine I am boiling water for tea. The knowledge that water boils at 100 °C is usually not conscious, but preconscious. However, it can be made conscious if this information becomes relevant.

[37]Consider when Sigmund Freud defined this terminology. Since then, research has come a long way. Concepts such as energy, to take another example, were given a clear mathematical-physical basis only through the work of Emmy Noether and Albert Einstein. And terms of engineering were not developed axiomatically yet. Maybe some mathematicians could imagine something with the terms data model and functional model, but hardly anyone else. This means that if psychoanalysis is to be officially understood as a natural science (Freud 2021b, pp. 439, 474)—and I assume that it cannot be otherwise—both conceptual worlds have to be adapted to each other. In SiMA we have taken the first step towards a psychoanalytic axiomatics with the definitions given here. Further steps must follow.

apparatus, there are conscious, preconscious and unconscious pieces of information. The claim by some scientists that his first topical and the second topical model contradict each other because it is not possible that two models exist in parallel at the same time, and that this contradiction has yet to be resolved, is false. Sigmund Freud was completely correct in developing both models, they are not in contradiction with each other, only the three terms used in the first topical model were an unfortunate choice in the sense of today's information theory: *conscious, preconscious* and *unconscious* are properties of information and not functional units.

Sigmund Freud's equation of the primary process with the unconscious and the secondary process with the preconscious and the conscious may need to be revised in light of new findings. According to Mark Solms (2013), some mental functions of the primary process may also contain conscious information, and some functions of the secondary process may process unconscious information. However, this point is still controversial, also within the npsa (Hartmann 2019).

So, what are the crucial characteristics of the primary and the secondary process? What are their key functions? First of all, we have to consider that the primary process is the gateway to all the information that humans receive through their sensory and physical perceptions via millions of sensors 24 h a day. The primary process must filter, prepare, modify, and forward all this data so that the secondary process, the planning functional unit, can focus on the essentials in an up-to-date and forward-looking manner, especially in the long term. Alexander R. Luria, Antonio R. Damasio and Mark Solms describe this very vividly in Luria (2001, p. 43), Damasio (2010, p. 58), Solms and Turnbull (2002, p. 84). Primary and secondary processes share their tasks, as already mentioned in List (2009, pp. 77ff). To put it in a nutshell, one can roughly say:

> On the basis of drive demands, the primary process prepares the information in an associated manner, filters it according to the pleasure-unpleasure principle, proposes options for reactions and actions, also according to internalized social rules, before the secondary process has to check what is effective, reasonable and reality related.

The secondary process considers short- and long-term goals and ultimately decides what to do next and in the future. And what is the basis for associations and decisions? *Experiences that are valuated with emotions and feelings.* Based on them and the acquired abilities, solutions are developed.

Imagine the following situation: You are in a workshop giving a speech. Someone in the back row is greedily eating his sandwich. Another person is sipping her coffee with relish. There is an interesting person sitting in the middle of all the other people with whom you would like to get in touch in some way. The topic to be presented is challenging. Many people are bored, you are hungry, and you have to go to the bathroom. And does the speech itself benefit you? Why go through all this trouble? But a good speech can only be successful when one is free from such thoughts and is alone in the world of the presented subject and lives only for it in that moment. Therefore, there must be functions in the psychic apparatus that help to concentrate on the speech and to suppress all other impulses, perceptions and associations as far as possible.

Another, completely different situation: You are lying on the grass in the countryside, listening only to the sounds of nature. You want to let your mind wander. You do not want to be disturbed by social factors. You want to enjoy the relaxed and quiet time by indulging in your daydreams. Which functions of the psychic apparatus make free wandering imagination possible, and which suppress it, and how?

Everyone can imagine ugly situations, situations connected with enormous pain, situations in which one is under the greatest hardship, but also situations that one no longer understand at all or, on the contrary, in which one feels the highest happiness and satisfaction. All this information has to be captured and processed by the primary process. What associations from past experiences are there? What solutions are there? What information should be passed on so as not to overload the secondary process? The secondary process must be able to focus on its most important task in the here and now and concentrate on the future. Based on which criteria, however, does the primary process determine which information to pass on?

I want to remind of the example of the last Sect. 5.2: Martin Pongratz's approach for solving the problem of catching thrown objects. Martin Pongratz applied for the first time the "human principle" for the problem of catching an object. Pedagogy and psychology teach us how the infant learns to catch. These experiences (that formed the basis of the ability of catching objects) are not stored in an objective manner, instead these experiences are as a rule *valuated* in principle (in SiMA, the term valuation is rigidly defined, psychoanalysis speaks of cathexis). In fact, everything is *valuated*: objects, persons, events, acts or processes, everything. There is no objective storing (i.e. storing without valuation). Through perceptions, these valuated pieces of information are associated. With the new perceptions and the input information of the drive representatives, they lead to something new, which is again valuated and thus stored. And all of this is done through a multi-level valuation scheme: Information arriving in the primary process is valuated with a *quota of affect*, then the *basic emotions*, the *extended emotions* and the *neutralized psychic energy*[38] take effect, which will be discussed in more detail in Chap. 6. And it is only in the secondary process that anything and everything is valuated with *feelings*, where the feelings result from the previous valuation variables. It should be noted that psychoanalysis does not yet take into account these strict distinctions, especially between emotions and feelings, which are indispensable in SiMA. If one wants to simulate the Ψ-organ, this differentiation of the valuation variables must be made.

Let us first take a closer look at the primary process. I start with the possible information flowing through the psychic apparatus. From this, inferences can be drawn, which functions the primary and later the secondary process must possess, in order to guarantee the behavior, which one expects by the information.

---

[38] Here the term "energy" appears in a non-physical sense, which provokes the association of physical laws and comparisons, which is highly problematic. For this reason in SiMA the concept that is known to psychoanalysts as "psychic energy" has been axiomatically defined as *psychic intensity* (see Sect. 6.3.2). In this way the psychoanalytic theory can be used without creating axiomatic contradictions.

Fig. 3.19 A detailed snippet of the block *psychic apparatus* from Fig. 3.18

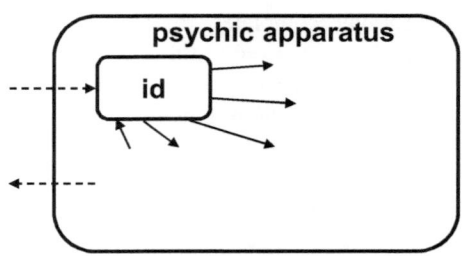

Let us begin with the *id*, one of the three functions defined in the second topical model of psychoanalysis, along with the *ego* and *superego*. According to Fig. 3.19 (a detailed section of Fig. 3.18), the inflowing, millionfold information from the neurological world first reaches the *id* (within the psychic apparatus) via an area not explored in neurology and psychoanalysis (in Figs. 3.18 and 3.19 in the area of the dashed lines). A distinction is made here between information of the *drives* and *perception*. The concept of *drive* is attributed to the body. The information of the drives is transformed in the psychic apparatus into *drive representatives*.[39] So, *drives* are information from the body about needs like hunger, thirst, sexuality, and so on. Their purpose is self-preservation or the preservation of the species. Each individual drive is assigned a *drive source, drive aim, drive object* and *quota of affect*.[40] The quota of affect represents a valuation by the respective drive representative. Is my hunger small or overpowering? The *id* perceives this unbalanced state of homeostasis in order to bring it back into balance. For this purpose, all drive representatives initiate associations that could lead to a solution. Associations are made with valuated objects, i.e. with quotas of affect, which promise a satisfaction of the drive representative.

The second major type of information that flows into the *id* is that of the many different perceptions of one's own body as well as all the perceptions that are registered outside of one's own body. They are used to release the inner drive tensions, or, to put it a little differently: They are adequately integrated into attempts at resolution of them. As long as it is information that is perceived from within, i.e. in one's own body, and concerns one's own body, recognition is generally less problematic for the psychic apparatus from an engineering point of view. One has "learned" one's body from the beginning and internalized it; one builds up an image

---

[39] From now on I will often speak of *drives* instead of *drive representatives*. This is a simplification I allow myself since it is clear what is meant and because in psychoanalysis the term drive representative is only seldomly used. In SiMA, however, we must make very precise differentiations, since both worlds, the world of hardware (to which the drives belong) and the world of the psychic apparatus (in which there are only representatives of drives), have to be treated in a totally different manner.

[40] Psychoanalysis generally speaks of affects, a term that is handled differently, thus in SiMA this information had to be defined more specifically (Dietrich et al. 2015). In SiMA, only the term *quota of affect* is used.

of one's own body schema in the *self*,[41] which is of course subject to change over the years. It is more difficult with objects that are perceived through the eyes, nose, ears or our haptic system. How does recognition occur here? When I see a strange, weird cup, I associate corresponding shapes, colors, maybe also the corresponding weight and my haptic perception. But what triggers the recognition that today's robots still have difficulties with? Psychoanalysis does not know an answer to this question.[42] In artificial intelligence it is called the *symbol grounding problem*. Following an invitation of mine Mark Solms (Dokaupil 2006b; Solms and Turnbull 2002, p. 279) explained to us that by valuating all objects the *self* establishes a relation to the object. It involves the aspect of *consciousness*. With a sentence I read, I do not only associate unconscious memories, but I associate an event to which I establish an unconscious *emotional relationship*. I thereby also develop a relationship to the content of the sentence, which can express itself in conscious feelings, just as it always happens towards all objects that I see consciously.

I am well aware of the fact that this presentation of the problem is not yet mature and certainly cannot be the final explanation for consciousness. In SiMA corresponding experiments are pending, hence this theory must still be thoroughly investigated and above all tested from our side. This is my apology for formulating the *symbol grounding problem* so vaguely. Brigitte Palensky was the first to address this issue in SiMA. Her results can be found in Palensky (2008).

In this context, I would like to address in this context how in the primary and secondary process information is processed and stored as symbols in general. In the primary process these symbols are *thing representatives*, in the secondary process *word representatives* (Freud 2020b, p. 298). The term *thing representatives* means that no words are assigned to them. When the thing representatives are passed on from the primary to the secondary process, they are assigned a word. All symbols, whether thing or word representatives, possess two crucial common properties: First, each symbol is embedded in one or more association symbol networks (Zeilinger 2010; Luria 2001, p. 73; Wilhelm 1982; Online Lexicon), and second, a symbol (whether associated via memory or perceived via sensors) is valuated via a quota of affect, emotion, and/or feeling (in psychoanalytic language: the symbol is *cathected*). Let us consider the example of my own coffee cup. It is characterized by a special symbol (thing representative) and has a high relation to my self (because it is my cup), so it has a high valuation. I have had it for a long time and it reminds me of many interesting situations. It is linked to many other symbols such as color, shape, weight, but also to symbols of other cups and equally heated debates. Thus, our knowledge and experiences are based on an enormous variety of networks and subnetworks of symbols of thing representatives, which are constantly growing throughout our lives. When I perceive something, stored thing representatives are

---

[41] Much attention will be paid to the self in the following. In psychoanalysis, it has not yet been dealt with extensively, which is why a separate research subproject was initiated for it in the SiMA project: "The self of a robot" (Doblhammer 2013).

[42] Psychoanalysis has to this day not seen this as an important question for clinical practice.

associated, and those that are most similar to them are marked as associated and thus *recognized*.[43] Thus, one can only recognize what one has already stored as thing representatives at least in a similar form and was able to integrate into the network of one's thing and word representatives. I prefer my coffee cup to a strange cup. Since in the primary process everything happens unconsciously, these processes cannot be measured and grasped from the outside, nor can they be reported from a subjective perspective (Freud 2020b). I can only deduce these relations from the model of psychoanalysis, which explores it and tests it via experiments.[44]

The networks of thing representatives and the processes derived from them in the primary process are not based on causal logic and real time connections. The thing representatives are linked exclusively associatively and thus also offer an enormously large space for fantasizing, i.e. to search for solutions, to balance homeostasis, to react to perceptions, etc., without having to be limited from the outset by annoying barriers of causality and real time constraints,[45] which govern the secondary process. One can understand the association processes of the primary process as extensive, unconscious *brainstorming*. Without the primary process the great realm of *fantasy* and all its feats would not be possible. *This is what distinguishes humans decisively from today's machines.*

It was perhaps noticed that the unambiguous assignment of functions to the primary or secondary process depending on whether they contain unconscious, preconscious or conscious information—as Sigmund Freud still assumed in his first topical model—is not possible. In the secondary process, there are some functions that exclusively process unconscious information.[46] For this reason, it was determined in SiMA that the distinction between primary and secondary process is made on the basis of whether the function works only with *thing representatives* or also with *word representatives*. Only functions that work with *word representatives* belong to the secondary process, all others to the primary process.

Let me return to the above mentioned example in order to perhaps make these considerations somewhat more tangible. I assume that some readers have

---

[43] This is axiomatically not correct, because in SiMA, recognition is attributed to conscious activity, that is, to the activity of the secondary process, when there is a valuation via emotions. Therefore, a new term must still be searched for the "recognition" in the primary process via quotas of affect and emotions, which has not been done so far.

[44] I define the term *experiment* in a broad sense. Psychoanalysts prefer to speak of clinical experience. But if one compares the psychoanalytic approach with those of natural sciences, then its cautious approach and its efforts to treat patients guided by theory are experiments in the natural scientific sense. In this sense, unconscious processes can be tapped from particular psychic material in a psychoanalysis together with the patient. Such a psychic material can be, for example, lapses of all kind, dreams, or psychopathological symptoms (Freud 2020d, pp. 206–210).

[45] This is an interesting aspect. So, if you remember a dream in which you fly over forests or go from one place to another, these dreams have already been reworked by the secondary process, because only through it and the linguistic recording of the dream material can the temporal aspect flow in, only it can create seemingly causal connections. They are just not perfect. They are very infantile.

[46] One example would be assessment whether a solution found for the next action can actually be carried out in reality. But this will be discussed in more detail in Chap. 6.

experienced this situation themselves. The reader may think of the example given above: One is giving a speech and has many needs to satisfy and corresponding active associations. But what is associated and what effect do the valuations (cathexes) of these associations cause? The *networks of associations* are interwoven in the way we perceive the thing representatives. One may find similarities but each person has in principle their *own individual network* of thing representatives. It is acquired. Externally perceived objects appeal to stored thing representatives, which with their valuations put a *valuation stamp* on what is perceived. If I find a certain type of cup or political party ugly, it will take some effort to dissuade me from this attitude. On the other hand, it must be clear that memories are changed by repeated recall and thus modified storage.[47] Statements like: "I remember exactly!" thus depend strongly on what has been associated and, again valuated, stored during multiple associations of these memories. Thus, the memory networks are subject to a dynamic change in the course of time, which offers disadvantages, but also advantages for the respective person. In any case, the change of the memories cannot be perceived, which is why one insists so often on the validity of memories, which cannot possibly be accurate.

The many associations that contain remembered proposals for solutions must already be filtered in the primary process in order not to flood the secondary process and force it to make too many decisions. The secondary process must always remain capable of acting in relation to the outside world and be able to concentrate on the essentials, while hunting, eating, loving, thinking or resting. Since the primary process only works with associations of thing representatives that were somehow connected in the past, without these connections being checked for causality, logical coherence and temporally meaningful sequences, the functions of the primary process develop the wildest and most absurd solutions. In addition, these solutions are influenced and cathected (valuated) by the current drive situation, so that drive-relevant elements are prioritized. The primary process must therefore carry out a more or less strict filtering in accordance with certain criteria via a *defense system*, which is capable of eliminating the majority of the proposed solutions or also transforming them into forms with which the person can live better. The regulatory factor for this process is that the solutions should help the person to gain pleasure in the real situation and, as far as possible, avoid unpleasure. It must be taken into account that pleasure and unpleasure are two independent variables.

Humans are forced to live and work with other humans. They need social rules, social directives. For this there is an instance in the psychic apparatus, the function *superego*. Proceeding from the problem that the many clashing information from the inner and the external worlds must always lead to conflicts, the demands and requirements of the superego naturally contribute additionally to conflicts. In order to solve these conflicts in such a way that one can live with them, the psychic apparatus has developed in the course of human evolution a relatively complicated

---

[47] Experiments in SiMA focusing on *learning* clearly show this (Fittner 2021).

but highly effective functional principle, the *defense system*, which is developed in different ways depending on the specific culture.

The defense system is based on numerous different principles. Research in this field is far from complete. But the scientific results obtained so far are difficult to comprehend for the non-expert. For reasons of time and effort, I can only go into a few of them here. The aim of the defense system is *to guarantee the highest possible pleasure and the lowest possible unpleasure for the own person in an inner-psychic conflict situation* under consideration of the momentary or future tasks which have to be solved. Valuated memories are used as a basis, which are weighed against perceptions. Experiences that were highly valuated in childhood play a decisive role in this unconscious search for solutions.

In order to make this difficult aspect transparent and to show what amazing feats the defense is capable of, ultimately with the goal of survival, I would like to start with a particularly impressive example. As a baby, one possesses the reflex to suckle at the mother's breast when hungry. This leads to corresponding symbols, which the baby learns. The *thing representatives* of *hunger* and *breast* are thus linked associatively from birth. In principle, when one of the two thing representatives is called up, the other is also associated with it. However, this would lead to irritation in an adolescent or adult person, which is why this connection is repressed, but already at the age of 1–2 years via a so-called *primal repression*. This is when an interesting process takes place: Initially, the nursing breast represents the greatest possible oral pleasure. However, if the breast is then withdrawn from the child when it is weaned, this represents a massive psychological burden, as it generates the highest possible unpleasure. The solution for the child is to find another object, a less conflictual substitute, even if this substitute via *displacement* does not bring the same satisfaction as the breast. The primal associative connection between breast and hunger must, of course, never become directly conscious. However, this goal is complicated by the fact that the storage of this chain of associations cannot be erased, since it is always activated again and again when hunger arises. This constant activation of the synapses prevents a "forgetting". *Primal repression* achieves that this traumatic association becomes and stays unconscious forever. From there onwards the breast becomes an associative element from where hunger can be displaced in the primary process onto other object representatives. In this way the trauma of losing the breast is overcome.

The primal repression of the primal trauma thus lays the foundation for future displacements to other substitute actions, in this way the scenic constellation of the trauma (in the example given, sucking on the breast) never leaves the primary process and thus always remains unconscious, which is of great benefit for the sanity of the individual.

The function of defense has developed many other mechanisms besides *displacement*: *denial, projection, repression* and some others. These different principles modify the incoming information or go as far as repressing or distorting it (up to the point of turning it into its opposite), thus helping the ego to choose a behavior that still offers a chance of gaining *pleasure* and to avoid *unpleasure*. And all these processes are continuously valuated by quotas of affect or emotions respectively.

In the course of the last hundred years, psychoanalysis has been able to explain many of the conceptions of defense addressed here. Taken together they present themselves as extremely complex, especially because they often occur in a combined form. Yet, basically all of them can be explained logically, which can be tested and evaluated by means of a *simulation*, as it is elaborated in SiMA.

Let me briefly introduce some defensive principles that are not too difficult to simulate in SiMA. I also want to explain in this way what you can imagine under these terms.

Denial

Me and my siblings asked our parents, when they were already a bit aged, what the one who outlives the other intends to do. Our mother first and spontaneously replied: "If one of us dies, I'll move in with my daughter." We laughed wholeheartedly, as it is a classic example of denial. Looked at closely, it is clear that my mother was blocking out her own death at that moment. Denial, of course, promised a greater gain in pleasure than the inclusion of the thought of one's own demise.

Projection

Baruch de Spinoza wrote that man created God in his own image (Mauthner 2015, pos. 410). Thus, humans project their wishes into a divine being, in which it is worth believing. Not for nothing famous scientists were enthusiastic about his logical conclusions, as I already mentioned. And if, as a small child, I imagined God with a white, curly beard as a kind, just father, in contrast to my father, who in my eyes was so often unjust and mean, what is that but projection[48]? Which promises more pleasure and less unpleasure? My father or God? This line of thought can be pursued further, which can be understood very well in Yalom (2012): Why do people like to join a religious community even though they may not really understand what is being preached? People grow up in a family and need a social community that promises them a certain warmth and support. And in this example you can see very well that defensive principles also overlap. Because on the one hand you project a lot into the religious community, into God, but on the other hand you also have to deny reality massively in order to be able to accept it as your own community, despite all the sometimes horrible abuses.

Regression

Regression is a falling back into an earlier stage of development. Many people regress in old age when facing their own death. But regression can also be observed in small children who get a younger sibling and are overwhelmed by the narcistic blow of no longer being the center of attention. Suddenly they want to drink from the bottle again, although they had already weaned themselves. So, they try to cope with the crisis by falling back into an earlier stage of development when they felt more cared for.

---

[48] This definition relies on colloquial language and understanding of the term projection. Sigmund Freud saw this more specifically. See appendix.

Repression

The defense mechanism of repression often characterizes the coping attempt. A simple example would be: An adult son forgets his mother's birthday, even though he has made up his mind to visit her. He is embarrassed. This mistake may be based on a repression of the son's aggressive tendencies toward his mother, which must not become conscious because otherwise he would be a bad son. The unconscious conflict between aggressive impulses and social demands on the mother, which would have come to a head during a visit, could be avoided for the time being by the slip of the memory.

But there are also more difficult and entrenched cases of displacement in adults who have experienced extreme trauma. They often suffer from depression or social phobia. Often, it is not until they go to therapy that they are able to come to terms with the fact that they were subjected to, for example, sexual assault when they were young. The repression then helps them to get over the tremendous pain that has been caused by the sadistic subjugation of their person by the perpetrator and the thus experienced helplessness.

Let us return to the large block mentioned above: the *superego*. Humans live, as mentioned, in social structures in which there are rules of behavior. "Be nice to each other", or "fear God", would be prominent examples. Through education or identification, these rules are more or less strictly internalized, that is, erected into the psychic apparatus as its own instance, the *superego*. Actions, initiated via the id, are always checked to what extent they violate these superego rules. Does the action plan have to be modified? Does it perhaps have to be discarded altogether? Can it be allowed? These are the questions that arise from the conflict situation and are ultimately resolved via the pleasure principle.

Psychoanalysis differentiates between the *reactive* and the *proactive function* of the superego. The reactive function of the superego works as follows. Any information that is to be transferred from the primary to the secondary process must pass through this function, where it is checked to see whether it violates internalized (social) rules. If this is the case, a conflict arises, which is processed by the defense system. It must be understood that this checking is, of course, not a simple yes-or-no decision. The passing information is then valuated by the quota of affect or emotions, as well as by the demands of the superego function activated by the passing information. The combination of all this information forms the basis for the decision of the reactive superego. The reactive superego thus basically reacts on the basis of a drive demand or a perception of the external world.

The proactive function of the superego, on the other hand, generates an action of the inner world that does not check information that stems from the id and is supposed to reach the secondary process but produces itself such information. As a striking example, I would like to recall a requirement of the Scout Movement: "Do a good deed every day!" With sufficient internalization, following this imperative promises pleasure, even though no external stimulus that actually demands the fulfillment of this requirement needs to be present.

In a healthy psychic apparatus, the functions defense and superego are attuned to each other. The few pieces of information which eventually pass the strict selection of defense are subsequently prepared for the secondary process by special functions (in SiMA we call them the transformation functions). Words must be assigned to the object representatives. Only representatives that are linked to word representatives can become conscious, which also explains why one cannot remember events that predate one's advent of word formation (Kaplan-Solms and Solms 2001, pp. 274–275). And another crucial task falls to these transformation functions: *Feelings* have to be determined from the various valuations such as *quotas of affect*, *basic emotions*, and *extended emotions*, because in the secondary process only the feelings as a unified valuation become conscious.

One of the first functions of the secondary process has the task of extracting possible goals for action from the information reaching it. This will, of course, depend on what my attention is focused on, and that in turn depends on where I am located. If I see myself as a speaker in a lecture hall, I will have different action goals in mind than if I am on a lonely hike through the plains. The information thus obtained must be subjected to a first reality check. Is the causality correct? Are the temporal relationships correct? Secondly, it has to be checked whether the conclusions are in accordance with my ideas of short- and long-term wish fulfillment. These are all processes that are still largely shaped by unconscious information.

Only now the information is sufficiently refined that the ego specifies via a decision function which goal it wants to pursue. As we will see later, the decisive influences are manifold. But how does the ego get from its decision, about which goal to pursue, to an actual action? On one of his visits (Dokaupil 2006a) Mark Solms explained to us that the ego generates several *imaginary actions* and plays them through. Thus, quasi-parallel "films" ("simulations", "map exercises", whatever you want to call it) run before each action, and the sequence that promises the greatest success is interpreted (this is linked to Sigmund Freud's conception that (Freud 2018b, p. 178) (freely quoting) "Thinking is . . . essentially a test action . . . ").

Thereby, a clear planning of the action with an included reality check as well as a valuation of the imaginary action takes place. This happens before the action itself flows into the motion control, i.e. before the decision is made which muscles are specifically controlled and how.

One last detail deserves to be mentioned in this chapter. It is obvious that the explanations of the secondary process, in contrast to those of the primary process, have been extremely brief. On the one hand, this has to do with the fact that psychoanalysis focuses on the primary process, which is also the central research goal of SiMA. One must first understand this domain before one can address the domain of the psychic apparatus that we are consciously aware of. Second, the area of conscious information is also not treated in such detail because the central topic of AI today is human behavior of that we are consciously aware of. In SiMA, on the other hand, I would like to enter the unknown realm of unconscious psychic processes.

Chapter 4
Why Do Humans Need Machines with Consciousness?

Why do humans need consciousness? Maybe one can infer from the answer why machines will have to have consciousness in the future. Yet, although I have provided a preliminary preparation in Sect. 2.3, I still need to define necessary terms—also in the context of SiMA—in order to be able to deal with this issue in a focused manner. This is because especially terms like *consciousness*, *complex process* or *function* are handled differently by scientists and particularly in common language.[1] This means that we cannot avoid a preliminary clarification of terms. And already the first subchapter (Braitenberg Vehicles—Complex Processes) will show that a somewhat more extensive explanation makes the problem better understandable. What does it mean not to understand the behavior of an object or a human being? *Why is then the conception of a functional model—like the ones that Alexander R. Luria and Sigmund Freud developed—useful to gain an understanding of the behavior of humans?* How can this be understood from the point of view of technology on the one hand and from the point of view of psychoanalysis on the other hand? What has to be considered with the term *intelligence*, a term which is avoided in psychoanalysis as far as possible, but which the field of AI carries in its very name? The insights gained from the work that has been conducted to answer these kinds of questions—which includes crucial insights from Alexander R. Luria, Antonio R. Damasio, and Mark Solms—will make it relatively easy to justify why AI cannot do without consciousness. *Consciousness cannot be suppressed in the realm of technology.* One must not shy away from dealing with it.

One must understand: Human beings are distinguished by their consciousness. Without consciousness you are not human. So, if we want to have an intelligence on the level of a human being, we have to deal with consciousness in detail. But let me first point to a simple thought experiment (Braitenberg Vehicles) that explains why the view of some scientists that certain insects must have consciousness (see for example (Internet: Spiegel online 2008) or (Tagesspiegel 2018)) is essentially

---

[1] And these definitions must be in clear harmony with all other terminological definitions of SiMA.

D. Dietrich, V. Hartmann Cardelle, *Simulating the Mind II*,
https://doi.org/10.1007/978-3-031-69530-8_4

unworthy of serious consideration. We have to learn to use more precise terms in this area.

And another important point. If you have ever tried to read *A Brief History of Time* by Stephen Hawking as a non-physicist, you will feel that you quickly reach your limits, especially if you are not familiar with the conceptual world of physics or have lost touch with it, as I have in recent years. Then you should first read *Hawking in the Nutshell* by Florian Freistetter (2019). He vividly demonstrated to me that theories that are essentially based on mathematical theories are not so easy to explain, or the explanations become fuzzy. Black holes and especially the Big Bang, Stephen Hawking's theory, often raise big questions of understanding. For example, people ask: How one can imagine the world before the Big Bang? Without the relevant mathematical and physical knowledge this cannot to be understood. However, if one deals with it, you get more and more of a feeling for it, and you start to understand it and find your way in this world of thought.

Nevertheless, in the following I do not need mathematical formalisms, but a certain mathematical way of thinking, which makes it possible for me to reach the necessary understanding. On this basis, I can build the bridge between physical and chemical laws on the one hand and the laws of the information theory on the other hand. And I believe that all intellectually interested people have the ability to achieve this understanding.

4.1 Braitenberg Vehicles: Complex Processes

There is an unwritten rule in computer technology and computer science that you should not base your work on literature that is more than 5–10 years old, with the exception of textbooks. I can largely agree with this, because these two fields change extremely quickly. Who cares about the architecture of a processor that is 10 years old? Every year, new generations of processors, storage devices, and software tools are introduced. In stark contrast to this rule, one of my colleagues pointed me to a 1984 book (Braitenberg 1984) that I was critical of at first, but which then surprised and fascinated me, even though it seemed so old and outdated at first. I am still fascinated by this book today, because it presents a solution, so to speak, of how a common conceptualization or axiomatics can take place for completely different fields, computer technology and psychology. And because it shows what the decisive step is to illustrate as simply as possible the distinction between a function and the structure of functions on the one hand and its behavior on the other. And it is mainly under this second aspect that I want to discuss Braitenberg's vehicles here. However, there is no denying that both aspects (the axiomatics and the distinction between function and behavior) have repeatedly caused problems in the SiMA project, which are always difficult to grasp for newcomers to the field.

Already the difference between the original English title of the book *Vehicles, Experiments in Synthetic Psychology* and the German translation (1993) *Vehikel, Experimente mit kybernetischen Wesen* (which could be retranslated as: *Vehicles,*

Experiments with Cybernetic Beings) is interesting. They seem to imply two completely different goals in terms of content and a different target audience. Why did they not stay closer to the original title? I can imagine that some readers had a hard time with the direct translation. What is synthetic psychology? Psychology is a man-made science. Therefore, such a distinction between synthetic and non-synthetic is not possible. The English title seems contradictory, but it appeals to people interested in psychology. In this book, Valentin Braitenberg wants to show that certain (artificially) developed vehicles exhibit behavior that we know from psychology and biology, where behavior is described in terms that one would hardly come up with in engineering. In other words, he shows through his experiments that vehicles based on an electronic control system can be constructed to exhibit certain behavioral phenomena known from animals in biology. Unfortunately, the purely technical formulation of the German title does not express this at all.[2] The English title, on the other hand, suggests such associations.

I would like to begin with a distinction that is fundamental to the description of the Ψ-organ and that Valentin Braitenberg introduces in a playful way: What is a *function* in the sense of electronics or computer technology,[3] and what is the corresponding *behavior*? Why do we need this distinction between *function* and *behavior*, and why is it so important? Valentin Braitenberg does not say all this directly, perhaps in order not to formulate it too technically, which might scare off psychologists. Thus, he explains this connection and the simultaneous differentiation in a self-evident way, and the reader, without perhaps realizing it, is gently drawn into this way of thinking. Valentin Braitenberg can do this because his primary goal is to show that psychology and modern technology can be brought together conceptually. But I want more. I want to show that—in order to be able to describe the Ψ-organ as a unit—the gap between the psychic apparatus and the physiological (anatomical) body can be bridged with the help of the information theory developed in computer technology. Therefore, I have to concentrate on working out this *bridge* in a precise way. In any case, I can recommend Valentin Braitenberg's book to everyone. It is fun to read and you get a different approach to the topic of *AI* in an accessible way.

In order to better understand Braitenberg's ideas, one has to keep in mind that his vehicles consist of four functions: a *mechanical body* of the object to be moved (chassis), the *drive unit* (e.g. a motor that drives the wheels of the mechanical body or a jet engine that moves the object), *sensors* that are used for perception (light,

---

[2] In general, I have the impression that the English version is more lively. The German version reads somewhat dry to me. Already on the cover of the English version there are imaginative satellites, on the German cover you see the wheels of a vehicle in the grass, which reminds me of a baby carriage. The aesthetic effect for me is less imaginative, less futuristic.

[3] The term *function* is used in different ways in science. In mathematics it represents the mathematical dependence of a value on its variables. How is this value calculated from the variables? In electronics and computer technology, a *function* represents the structure of electronic components that change the information passing through them until it is finally available at the output as an output variable. The electronic components are again to be regarded as sub-*functions*.

Fig. 4.1 Second experiment with developed vehicles

smell, color, etc.), and an *information unit* (in this case, of course, an electronic unit) that is responsible for the automatic control of the vehicle. Valentin Braitenberg thus deals with two "worlds",[4] the physical world and the world in which only the laws of information processing apply. The sensors (in biology and medicine they are called receptors) and the drive units (in engineering they are called actuators) form the interface between the two worlds. Let us look at this in a less abstract and dry way and try to see it as vividly as Valentin Braitenberg himself does.

It begins with a simple mental experiment. A very simple vehicle like a *scooter*, in which one wheel is driven by a motor that is connected by a cable (and some basic electronics) to a light sensor. The more light the sensor receives, the faster the motorized wheel of the scooter turns. So, one function is the drive unit for that wheel, another is the sensor, and a third function would be the electronics, which is the informational link between the sensor and the drive. All three functions are interrelated, which we call the *structure* (of the functions). *The functions and their structure determine the behavior of the object (the scooter)*. In this case they are very simple and can be described in a simple way.[5]

The second experiment is more complicated. The *sensor* and *drive functions* are coupled and initially controlled electronically, as shown on the left in Fig. 4.1. The plus signs mean, as in the first experiment, that the wheel turns faster with increasing light input (luminous intensity) to the corresponding sensor. The left sensor of the left vehicle is further away from the light and therefore receives less light energy, so the left wheel turns slower than the right one. The vehicle will therefore turn away from the light. The opposite behavior is observed for the right vehicle in Fig. 4.1, although it contains the same functions. However, the structure of the functions is different. The right sensor communicates with the left wheel drive. Because this right sensor receives a weaker luminous intensity than the left one, the vehicle moves toward the light.

---

[4]Volker Hartmann Cardelle pointed out to me that I did not formulate this clearly and it can lead to a misunderstanding. Therefore, here my explicit note: If I use the term "world", I mean this in the transferred sense (epistemologically, in the logical sense, concerning a "world of thought") and not ontologically.

[5]In the sense of a hierarchy of functions, the *scooter* as a whole is also to be seen as a *function* that is hierarchically arranged above the other functions such as the drive unit, sensor, etc.

Fig. 4.2 Different relations between luminous intensity and velocity of the drive units

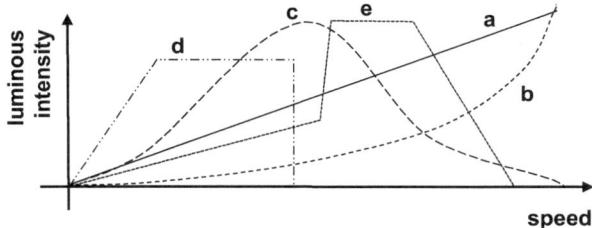

Again, the structure of the functions (i.e. how they are related to each other) and the resulting behavior can be described in a simple way.

It becomes more complicated when the relationship between luminous intensity and wheel speed is no longer linear (like in Fig. 4.2, curve (a)), but exponential as in curve (b), or nonlinear as in curve (c), or even nonlinear and discontinuous as in curves (d) and (e).

For example, in the spirit of Valentin Braitenberg, one could describe the behavior of the vehicle, assuming curve (c), *psychologically* in such a way that the vehicle aggressively increases its speed as the luminous intensity falling on the sensor increases, until it reaches a maximum at which the aggression turns into fear, and the vehicle loses more and more speed until it stops completely at a certain luminous intensity because the fear of the luminous intensity becomes too great.

Even if such behavior patterns can be interpreted into the seemingly strangely moving vehicles, such interpretations are nonsense. There is no *fear* or *aggression* behind such behavior in this experiment. The connection between function and behavior, i.e. the interaction of drive, sensor and electronic functions and the resulting behavior, can be traced back to simple natural scientific laws. There is no hardware or software function in the vehicle that produces aggression[6] or fear. Aggression or fear are emotions in animals and humans, which, according to Antonio R. Damasio (2003, p. 48), require a certain intelligence, which in turn must have a minimum number of nerves as well as a corresponding nerve structure of the Ψ-organ (here a corresponding electronic system), which was discussed in Chap. 3. The behavior of these vehicles is thus generated by a simple electronic hardware, and since we develop it ourselves as engineers, we also know the relationships between the functions and their structure and the resulting behavior, and can both fully describe and simulate them. Valentin Braitenberg wants to point out, among other things, that one has to be careful when using psychological terms, partly because one might assume too complicated functions and structures in the dynamic objects, and partly because such terms by definition require functions and structures that the object does not even possess.

---

[6]I call the attributions of humanoid behavior patterns a projection of oneself onto the machine, which means that we perceive something as being human-like, but this perception mostly relies on our expectation and not on the inner structure of the object in question, thus what we perceive to a large extent comes from our psychic apparatus and not the external world.

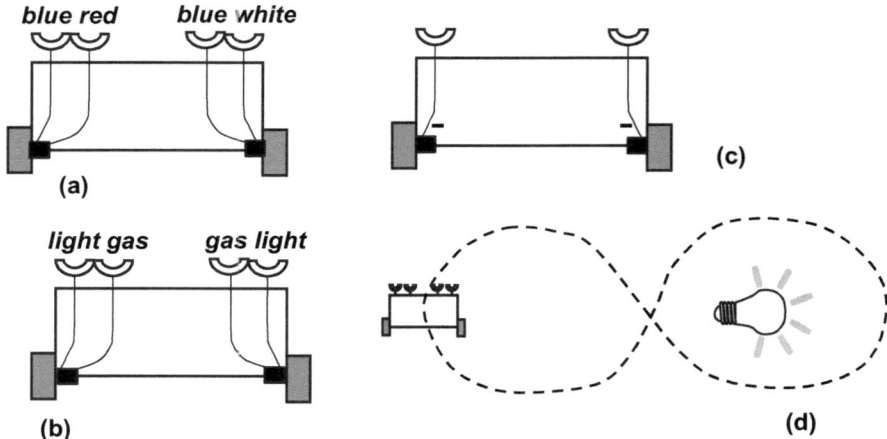

Fig. 4.3 Modifications of the functionalities and changes of the structure of the functions can lead to strange behavior

Yet another aspect, which I already mentioned in Sect. 3.1, becomes apparent here. The complex behavior of bees or ants does not indicate a high intelligence that could be compared to human intelligence. Even the simplest neuronal circuits or simple electronics can produce complex behavior, which can only be explained functionally, i.e. by the functions of the system and its structure.

In his further experiments, Valentin Braitenberg integrates more and more functions, modifies them by designing sensors for certain colors (Fig. 4.3a), by introducing different gas sensors (Fig. 4.3b), or by complicating them, for example by determining that increasing light intensity does not lead to an increase in speed, but to a decrease in speed (Fig. 4.3c), etc., and of course he varies the structure of the functions in different ways. The vehicles exhibit a behavior that is increasingly difficult to understand in spite of the simple electronics behind them, i.e. in spite of the relatively simple functions and the relatively simple structure of the functions. Vehicles that follow a lying eight is in this context still a relatively simple process. The application of the different curves in Fig. 4.2, the different modifications of the functions and their structures make the behavior more and more opaque. Those who do not know the functional design and the functional structure will not be able to understand nor to describe the entire behavior of a complicated functional structure. Only certain isolated behavioral phenomena can be described. This is illustrated in Fig. 4.4: At point L (limit), the work required to describe the *dynamic behavior* goes to infinity, but the *functions* and their *structure* can still be described in a relatively simple way.

The object, when its behavior can no longer be described, might appear to be "alive" if one did not know that it consists essentially only of sheet metal and silicon. It is important to keep in mind that research in biology and psychoanalysis is different from research in computer technology. Perhaps, in a sense, the approaches

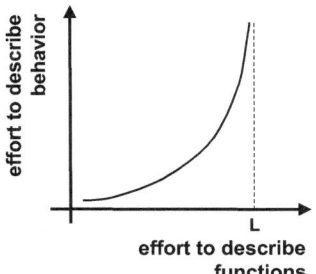

functions: *complicated, but can still be described*

behavior: *complicated to complex; if complex, they are no longer completely describable, but only individual phenomena of the entire behavior*

Fig. 4.4 Definition of the complex domain. L: Limit

can even be seen as opposites. A computer engineer designs the functions and structure of the computer to be built based on a behavioral specification that a customer predetermines for a process. After simulating, testing, and verifying the model of the computer, the computer is realized, and all that remains is to verify that it exhibits the behavior specified (by the customer). Biologists and psychoanalysts, on the other hand, have to work out the functions and their structure from the behavior of animals and humans, which already in the case of hardware (i.e. physiology, the nervous system of the Ψ-organ) often raises difficult questions, but in the mental sphere must be considered extremely complex. The results of research on the psychic apparatus could thus far not be functionally simulated and therefore not be experimentally (functionally) tested (statistical methods must be excluded, I will talk about them in Sect. 5.1). The question is, how do we arrive at functional models in neurobiology and neurology? Behavior always explains only phenomena, not the psychic apparatus as a function, as a whole system. And in the natural sciences, you look for functional models that show a certain behavior in experiments, so you can evaluate them. If we are talking about a few neurons in biological bodies, as in Bose (2010), where the number of neurons is in the low double digits, this is now possible with reasonable effort (I am talking about a simulation of the functions and their structure, not the behavior). If, however, the number of nerves in the biological body is significantly higher, then completely different methods must be used. And it is these methods that I want to explain in this book, since this was the task set for SiMA.

How does one usually proceed in psychology, psychoanalysis or biology in general? From the point where behavior can no longer be completely recorded and described (in an animal or in a human), one too often concentrates solely on particular phenomena of behavior. How, on the other hand, is one to arrive at the *definition of functions*? It is "relatively easy" when we look at the physiology of a biological body. The individual *functions* of the heart, stomach, or eye can be identified materially and observed directly in the body. Their *behavior*, i.e. how they work individually and interdependently, could obviously be worked out fairly well, even if we do not yet know all the details. It is much more difficult with the mental part. The functions of the psychic apparatus cannot be grasped (identified materially), they can only be *defined* with great effort and enormous prior knowledge. This was and is a decisive reason why SiMA concentrates on *functionally oriented psychoanalysis*. Here the expert develops the functions of the psychic apparatus on the basis of analyses of human behavior. The crucial point is to understand the flow of information through the functions, which makes it possible to obtain the structure of the functions. The functions and their structure are the hypothesis in the description of the psychic apparatus. They have to be evaluated and modified until they are in agreement with the results of the experiments.[7] This is the natural scientific procedure, as it is used, for example, in computer technology—my scientific field. Sigmund Freud also followed this idea, and as a result he had to constantly correct his results based on new findings. So the methods of us computer scientists are no different from those of functionally oriented psychoanalysts. In contrast psychologists and psychoanalysts who rely solely on attachment or object relations theories are generally limited to the study of behavioral phenomena. *They cannot arrive at a functional description of the psychic apparatus as a whole.* From the point of view of the natural sciences, however, *functional description is the prerequisite for being able to explain all perceptible behavioral phenomena.*

Behind these considerations lies a scientific problem that may require some intellectual acrobatics on the part of the reader. But the scientific depth and the knowledge of the underlying design and abstraction problems help to understand the problem better.

In the past, when computer engineers or computer scientists received the behavioral description of a control system from customers, it was very difficult to develop the functions and their structures based on this description. Let us take the example of a printing press. The customer, a manufacturer of printing presses, explains how the press should behave. Based on the customer's behavioral description, the computer designer must work out what functions are necessary to ensure that the motors, actuators, etc. of the printing press are correctly controlled. Based on these functions, the structure of the microcomputer can be calculated in detail until the electronic component, the chip, that controls the press can be produced. In the past, this path of developing the functions of the electronic circuit was mandatory and a challenge for many engineers. In the meantime, tools (design languages) have been developed that make it much easier to develop the electronic chips with their functions and structure directly from the behavioral specification of a process. The prerequisite for this is that the behavioral description can be formulated in a sufficiently precise manner by means of the necessary abstraction and that the behavioral model does not deviate too much from the real process. In other words, the abstraction must not be too coarsely "carved out" in comparison to the real, dynamic process. This is where you may come up against the limits of what is

---

[7] In order to conduct experiments of the psychic apparatus on human beings, and thus develop the sub-functions, sub-sub-functions etc. of the ego, superego and id, the psychoanalyst needs simulation tools like that of SiMA. A decisive reason why Sigmund Freud was not able to work out a complete model of these sub-functions.

technically feasible. Nevertheless, this approach is useful for today's chip designers, because they know how the realized computer will behave on the basis of behavioral simulations. Such a product can be sold without worries. If, on the other hand, one takes the (from today's point of view) cumbersome path of developing functions and their structures, and first compiles the structure with the associated functions (which only a few students learn today), i.e. simulates this construct of individual functions, one can never be sure (unless simple, linear functions and a simple structure are the basis) that the chip will not exhibit unpredictable, *additional* behavior in certain situations. Therefore, before such a chip is sold, it must be extensively tested for *all possible* behaviors, not just the desired ones. And that means more effort and higher costs. For this reason, tools have been developed that automatically generate the functions based on a *complete behavioral* description.

In the case of the simulation of the Ψ-organ this is different. If one were to proceed in the same way with the psychic apparatus of the human being and to simulate its behavior, one would get as *artificial human being* what one has originally described, defined, specified and thus programmed as its *behavior*. Nothing more than that. However, the behavior of an individual human being is complex. It cannot be completely described, only characteristic phenomena. *If, however, one wants to understand the human being as a whole—i.e. all behavioral phenomena— there is no way around working out its functions and their structure.* And this is exactly what psychoanalysis aims for. But this brings us to the question of what *functions* actually are: They are the *physiological* and *psychic* components *that do something with the information*. They receive them at their respective input, modify, partially store and finally feed them to their output.

The functions developed in this way and their structure are therefore abstractions. Sigmund Freud developed the three functions *ego*, *id* and *superego* as main functions of the psychic apparatus. In SiMA, we derived their sub-functions from the theories of psychoanalysis (i.e. we subdivided the three functions into smaller sub-functions), and these in turn into sub-sub-functions, and so on, until we had such small functional structures that we could describe them with sufficient precision and ascribe to them a defined, desired behavior, which will be discussed in Chap. 6. The resulting structure with its numerous functions ultimately ensures that any kind of behavior of the object *human* can be simulated.[8] In contrast, if only a limited number of behavioral phenomena are specified and functions are thus defined, a much more restricted model is obtained. Robotics and AI apply this kind of models all the time. A famous example is shown by Breazeal (2002). It has nothing to do with a functional model of the Ψ-organ. How do you combine all the findings of

---

[8] You thus get a natural scientific model. However, scientific models, unlike mathematical models, are not provable. They always contain (unlike mathematical models) a residual inaccuracy that cannot be measured. No aircraft manufacturer will be able to *prove* that their aircraft will behave as they hope. Therefore, he must perform enough experiments to make the probability of a possible error as low as possible. This is how an aircraft designer *evaluates* and *validates* their designs. To put it more simply: The proof at purely mathematically imagined triangles is possible. A physically realized triangle will always have some inaccuracy. It can only be evaluated and validated.

psychology, developed by different schools, i.e. with different methods and theoretical foundations (thus lacking interoperability), without a functional, unified concept? How to unite physics, based on a behavioral description, with the psychic apparatus? And how do you take into account in such a behavioral model the memories of the human being, which influence in many different ways all kinds of functional units of the psychic apparatus (e.g. perception)? The more one goes into the relevant details, the more everything speaks against the approach via the simulation of a behavioral model. *From today's point of view, the simulation of the functional model can be the only useful solution.*

Back to Valentin Braitenberg. There is still a decisive step to be taken. Let us return to Fig. 4.4. We are talking here about objects to which a certain intelligence is attributed, i.e. in which information is processed. Let us assume that we are on the right side of point L (limit), from where on not all behavioral possibilities but only single behavioral phenomena of such an object can be described. In contrast, the functions of the object can still be described, but the description becomes more complicated and costly. But it is possible, as we can see in the example of the vehicle. Therefore, in the SiMA project we call the area to the right of L *complex* in contrast to the area to the left of L, which is defined as *non-complex* (it is "only" complicated).

This means that in *non-complex processes* the behavior may be complicated to describe, but the comprehension and description of this behavior, also via mathematical formalisms, is still feasible. Feasible means that the models of functions and their structure obtained by abstractions still exhibit behavior that can be captured and described in its entirety. These models include all behavioral phenomena of the object.

In the area of complex processes (i.e. to the right of L in Fig. 4.4), however, not all behavioral phenomena are known. Information about the process is missing and cannot be acquired. By means of experiments and analyses, functions with their structure (i.e. the functional model of the object obtained in this way) can be hypothetically worked out, but they may not cover the entire behavior of the real object. In such a case, computer engineers are accustomed to studying (analyzing) the obtained model through experiments and simulations—thus obtaining better and better refinement and more detailed description of the functions and the structure of the object—until one has worked out in sufficient detail all the behavioral phenomena that are important for the task in question. Important here means that the deviations from the desired behavior analyzed in experiments are not considered significant in the real process.

I would like to draw two interesting and important conclusions from this insight. First, processes can exhibit behavior that seems inexplicable at first. Especially if they are complex processes according to our definition. However, if one knows the functions of the hardware and/or the functions, simple (or at least understandable) explanations can emerge. Mystical explanations are always out of place. But how are these functions to be worked out? The experts—and in the case of SiMA they are psychoanalysts—can usually only do this by interpreting the behavior. This means, however, that the functions worked out in this way must always be tested to see if they stand up to further experimentation.

Second, complex processes remain complex only as long as not all known behavioral phenomena can be considered in the functional model. If we think of the process *human*, we can say with great certainty that it will be a never-ending story to explore its neurological and mental functions with increasing precision.

4.2   Intelligence

With consciousness comes the concept of intelligence. How can intelligence be defined? Various intelligence tests have been developed over the past few decades. What do they measure? Why does psychoanalysis avoid this term, while AI has it in its very name?

Intelligence is not a relevant issue in SiMA. I see this in a similar way to some psychoanalysts. In order to do psychoanalysis with a person, the person has to be able to reflect. The consequence is that they have to have a certain intelligence to be able to work on themselves. A certain intelligence is simply a basic requirement for collaborative work and social interaction. But how can it be measured or determined? Today, at my age, I allow myself to tell the following story at appropriate moments. In school, I was generally very diligent and ambitious. However, my grades were so poor that the school was forced to dismiss me early. My parents wanted to help me and referred me to the employment office, where they advised me to take an intelligence and ability test. What would be the best profession for me? The test was administered by three psychologists. Three young ladies of about the same age, about 18, were tested with me. The test lasted several hours, and a few days later I received a letter advising me *not* to become an electrician, even though it was my "dream job". They suggested that I apply to the railroad to become a ticket agent. My intelligence quotient was not sufficient for the craftsmanship of an electrician. I am still grateful to my father for strongly advising me to try what I desired to do. Everything else would take care of itself. He, too, had dropped out of school as a teenager and later managed to become a well-paid union leader. Then I did my apprenticeship as an electrician, which I finished with honors. Later I was also tested for intelligence by psychoanalysts at the Sigmund Freud Institute in Frankfurt. The results were very positive. And about a year later, I was able to graduate from high school as one of the best students in Rhineland-Palatinate and received an award for this achievement. What did the psychologists at the employment office measure at that time, and what did the psychoanalysts in Frankfurt determine? I don't know. I understood only one thing at that time: The concept of intelligence is not easy to handle.

Another story: When microcomputer technology was in its infancy, people wanted to know which computer was the most intelligent. The engineers started by comparing the MIPS (million instructions per second) and FLOPS (floating point operations per second) of different computers, but quickly realized that they were basing their evaluation on an invalid definition of *intelligence*. The number of hardware components of a computer and its computing speed can only be a measure of a minimum number of components and functional units necessary to achieve a certain "computer intelligence". It was understood that the *intelligence* of computers could only be determined in relation to their *application*. Then a distinction was made between hardware and software benchmarks, and finally specific tasks, task objectives, and experiments were defined against which the performance (= intelligence) of computers could be compared. Thus, it was understood that the statement that computer X is more intelligent than computer Y is nonsensical. *The term intelligence cannot be used in a value free manner, but must always be seen in the context of the tasks and their goals, and the corresponding tests. Intelligence needs a point of reference.*

For this reason I refuse to speak (Dietrich et al. 2017a; Apprich 2019) of a person who is "intelligent per se". Intelligence has to be related to a reference. There are the integrated circuits (ICs), the current basis of computers, which are able to solve certain automation tasks the fastest, the computer which is best suited for chess, or the computer which is able to answer knowledge questions the fastest. The same principle must apply to humans. One must always ask: What intelligence is needed for what task? Albert Einstein must be counted among the most intelligent people to solve physical problems. Whether his political actions were always very intelligent may be doubted, considering his changing attitudes toward the atomic bomb. Trump can be considered intelligent in his abilities as a real estate broker. Socially, morally and politically I doubt his abilities very much, especially when it comes to the way he treats people.

In Sect. 3.1 I talked about the fact that in SiMA we do not consider the structural or swarm intelligence, but only the intelligence that is *actively* performed by the *neuronal system* with its numerous functional units. In Sect. 3.2, it became clear that the amount of information flowing into the Ψ-organ at any given moment is enormous,[9] even in a moment of relaxation, e.g. during sleep. This makes it clear that intelligence must be related to the number of neurons and synapses. There can be no doubt about this, even if it seems to contradict what has been said above.

Having this in mind, let us come back to consciousness.[10] Consciousness necessitates a *self*. But the *self* presupposes that the human being has a very powerful information processing system with an enormous amount of stored information in the background. This information processing system is constantly confronted with a flood of information streaming into it. All of this requires a large number of neurons

---

[9] I learned about the continuous, strong activity of neurons mainly in the course of brain experiments of anesthetized cats with a stereotactic device (Meyer-Waarden et al. 1975). The neuronal impulses were amplified by loudspeakers that sounded like machine-gun bursts, although the cats were always anesthetized and relaxed on the operating table. Today, I no longer need to be convinced of the futility of most of these experiments, which I witnessed as a student at that time.

[10] Note that I speak about consciousness in the sense of the sum of our thoughts about the situation in which we are, which also includes our thoughts about the fact that we are thinking etc. I do not mean consciousness in the sense of being awake in contrast to being unconscious or passed out.

and synapses, which worms and flies do not have in such enormous numbers. This leads to the following conclusion: *The self requires a certain minimum number of neurons and synapses for consciousness to arise. But intelligence cannot be understood simply by counting the number of neurons and synapses. Intelligence must be related to the ability to process all information in the individual functions of the psychic apparatus according to specific tasks.*

This brings me back to what was said at the beginning of this chapter. Intelligence is not the subject of SiMA, because otherwise we would have to determine which intelligence we are talking about in relation to which reference. However, we are not interested in developing a mature model of the Ψ-organ. This cannot be the goal at the beginning of such a development. We want to show that the Ψ-organ can be modeled *in principle*, or that its *functional structure* can be worked out *in principle*, in order to substantiate with simulation *experiments* the findings of neuroscientists and psychoanalysts. And, in a further step, to incorporate these findings into AI research. Later on, it will be very well possible to conduct intelligence tests for the functional units elaborated in Chap. 6 (there are currently almost 50 functional units), or even to define and run various reference tasks for the entire model. But this should be left to future researchers. Then, however, one must also consider how exactly to define and distinguish the terms intelligence, performance, etc., and what the goal of such investigations and experiments should be. In the course of the history of computer technology, we had to learn a lot in this respect.

4.3 Consciousness and the Need for Axiomatic Definitions

The knowledge of neurology, neurobiology, psychology and psychoanalysis has increased considerably in the last decades. However, one topic has not yet been thoroughly explored in a scientific way: consciousness. Many scientists are convinced that the building blocks of consciousness must first be gathered piece by piece (Kaplan-Solms and Solms 2001, pp. 216–217). But does this mean that the individual neurological and mental functional units of the Ψ-organ must first be explored piece by piece? Is this even possible?

Others think that consciousness goes along with subjectivity, and in this case the crucial scientific principle speaks against dedicating oneself to this topic as a scientist: Science should only deal with entities that can be objectively measured. But should not this traditional way of thinking—which, after all, has developed through observation—raise doubts? *Consciousness is what defines a human being.* It is part of the human being and therefore part of the biological body. *For this reason, natural science must deal with it and study it. It is only necessary to find out and reflect on the fallacy of this seeming contradiction.*

A third serious argument is the replicability of experiments. Yes, it is true that a human being learns with almost every action, stores new information, and thus changes the boundary conditions for the next action. Repeatable experiments—which are necessary to verify the results—thus seem to be impossible with humans.

And only with the help of statistics could one come close to something like a repeatable experiment. However, it is an important requirement of natural science to repeat experiments with functional models—and these are assumed in SiMA—in order to verify previous results. How can this apparent contradiction be resolved?

Let us start with the simplest of the three problems mentioned, the demand that one must first work out the building blocks of consciousness. From my point of view, i.e. that of a computer engineer, this is fundamentally the wrong approach. In this case one would follow the bottom-up method (as we called it before), which I already had to reject in Sect. 3.1 as unacceptable for the modeling of intelligent systems. For an assembly of functions, i.e. the development of a functional model, one does not know with the bottom-up method whether an optimal solution, or any solution at all, can be found in this way. For the computer engineer (and in general also for the computer scientist) it is therefore a must to follow the top-down method. It would be wrong to start building a model in the neuronal domain, or even to try to work out the lower functions of the mental domain, before the "top" has been defined. And the bottom-up method is already out of the question because the space between the functions in which unconscious information is processed and the space of neurology is not directly observable for science.[11] Functions in this unknown area can in no way be developed experimentally from "below". How to get this knowledge? Here other methods have to be used, which have been proven in computer technology, as the development of computers has shown us for decades. However, I can only explain this in more detail and offer a corresponding solution to this question in Chaps. 5 and 6.

The second objection, that *consciousness is always subjective and that dealing with it therefore violates the principle of natural science*, is based on a belief that was (rightly) born at a time when people were just beginning to understand scientific work. And when I examine a dissertation today, it is still a big stumbling block for everyone, at least at the beginning. Again and again, unverified or even unverifiable claims are put into scientific contributions. But they have no place there. This is a challenge for almost every young scientist. Science must work *objectively* and remain verifiable in every respect. This is true, but this statement has absolutely nothing to do with the argument about psychic *subjectivity*. *Psychic subjectivity can very well be worked out objectively and in a scientific w*ay, one just has to learn how to do it. However, this is difficult. One needs patience with oneself, perhaps because one is not used to it in this area. You have to be very careful not to let subjective information creep in, which inevitably leads to errors. And my experience in SiMA so far is that many scientists find this difficult, very difficult indeed. Not less so for me! Although I have been working intensively on this topic for about 20 years, I still make mistakes, which is why I cannot avoid having my statements checked by other people.

The *problem of repeatability*: It is true that a human being—if you look at the details—will be in a different state at every moment. If you look at the states of all

---

[11] This is discussed in detail in Chap. 6.

the organs and functional units of the Ψ-organ at a given moment—no matter which one—the same state will never occur again. The human being is far too complex and is constantly changing and learning. However, SiMA is not about simulating Mr. Johnson or Mrs. Jones at a certain point in time, but about using simulations to demonstrate, on a natural scientific basis, whether the models developed by neurology and psychoanalysis are correct. It is about providing neurologists and psychoanalysts with tools to more easily test their hypotheses and models experimentally on a natural scientific basis. It is also about ensuring that the models obtained in this way can be used in artificial intelligence. In this sense, of course, all simulation experiments are reproducible in all details. I will talk about the necessary boundary conditions in Chap. 6.

If we want to model scientifically the human information system, from its neurons up to the psychic apparatus, and that is what we want, then the realm of consciousness has to be addressed as well. We cannot use any excuse to avoid this subject. Throughout human history there have been subjects that we were not even allowed to think about for various moral, religious, or power-political reasons, before someone did it, in the end always for the benefit of humanity: Dissecting human bodies was forbidden under penalty of death, but Aristotle did it. Nicolaus Copernicus declared that the Earth could not be at the center of our universe. The idea that man is not descended from Adam and Eve is still an outrage to many people today. And it is not for nothing that the teachings of Charles Darwin are still banned in some places. Let us dare to look at the subject of consciousness from a natural scientific point of view, even if many of my colleagues call people like me mystics. I am an engineer and I do not think I have ever deviated in any way from the basic attitude of a natural scientist. Bertrand Russel and John C. Eccles were and still are my role models. Antonio R. Damasio (1999, p. 82) also formulates a stance with which I can identify: "We scientists make it easy for ourselves when we lament the circumstance that consciousness is a completely personal and private matter and not accessible to that objective observation which is common, for example, in physics or in other fields of the life sciences. However, we must come to terms with this situation and make a virtue out of the burden. Above all, we must not attempt to study consciousness exclusively from an external perspective for fear that the internal perspective may be completely distorted. The study of human consciousness requires as much the inner as the outer point of view."

What has science developed so far on the subject of consciousness? Philosophical statements are of no help to us in the SiMA project, because we want to implement the model to be developed in an information-technical, natural-scientific, i.e. also practical way. For me as an engineer, philosophy is important in this area to avoid logical detours and to prevent errors from creeping into the model that would complicate our work. However, we technicians and engineers ourselves are even less able to explain what consciousness is, because this is not what we have worked out in our studies and in the course of our lives. As I said at the beginning, we have to refer to those who have been studying this subject from a scientific point of view for decades, and these are the neuroscientists, psychologists, and psychoanalysts.

And what do these scientists say on this topic?

Antonio R. Damasio (1999, p. 5) formulated the following statement about consciousness: "Consciousness is, in effect, the key to a life examined, for better and for worse, our beginner's permit into knowing all about the hunger, the thirst, the sex, the tears, the laughter, the kicks, the punches, the flow of images we call thought, the feelings, the words, the stories, the beliefs, the music and the poetry, the happiness and the ecstasy.". He went on to write that consciousness develops an interest in the self and helps us to refine the art of living. But what does that mean in practical terms? We cannot measure consciousness. Many try to define consciousness in terms of intelligent actions (Spiegel online 2008; Tagesspiegel 2018; Spektrum 2018; Frontiers 2018), or, even more simply, some attribute mind, consciousness, thought processes, and related abilities to humans alone, in line with Cartesian dualism. But does this fit with Darwinian thinking? Mark Solms and Antonio R. Damasio (Solms and Turnbull 2002, S. 50ff; Damasio 1999, p. 231) explain these different conceptions very well in their elaborations on the subject of consciousness. This shows that generally accepted (scientific) definitions of consciousness do not (yet) exist. Interestingly, the soul has been discussed since ancient times, and some even see a connection between the soul and consciousness (Hinterhuber 2001, p. 159), but consciousness as we know it today, only really came into the focus of interest in the twentieth century (Damasio 1999, p. 231). Christian religion scholars, on the other hand, are very interested in this topic, as Bauer (religion scholar, Münster) tries to explain in a convincing way. In this sense, he also addresses the issue of axiomatics and ambiguity, which is essential for SiMA. He argues (Bauer 2018, p. 15) that, on the one hand, people in our complex world ". . . tend to be intolerant of ambiguity, . . ." i.e. they avoid ". . . ambiguous, unclear, vague, contradictory situations, . . ." but, on the other hand, ambiguity is ". . . the only force that limits and de-emphasizes the destructive, genocidal potential of modernity". Bauer also emphasizes (Bauer 2018, p. 34) ". . . that religion is first and foremost communication . . .", in SiMA's sense, therefore a matter of the psychic apparatus and thus of consciousness and logic.

Philosophers also take up this topic again and again. However, these are all hermeneutic or purely logical model approaches, which can be overthrown tomorrow with a different logic, also depending on the spirit of the times. In SiMA, on the other hand, we exclusively follow the scientific approach, because only in this way is a functional simulation of the Ψ-organ possible, only in this way do we have a clear axiomatics, and only through such an experimental approach are repeatable experiments guaranteed. This is a completely different goal from that of religion and philosophy. It must be taken into account that the Ψ-organ is to be seen as an organ (Solms and Turnbull 2002, p. 8), which can only be understood in connection with the other physiological organs of humans (Solms and Turnbull 2002, p. 76). Thus, modeling the Ψ-organ without its physiological body has nothing to do with a biological Ψ-organ or a biological information system. All models and algorithms of AI, which are not based on a physical or physiological body, can only in a (very) limited sense have something to do with the way of thinking of humans and thus with their intelligence (Solms and Turnbull 2002, p. 18). *A simulation of the Ψ-organ in order to understand its structures, without considering the physiological* *body (in at least some way), is therefore nonsense.* The consequence is that the development of a functional model of the Ψ-organ, which presupposes the neurological *and* psychical functional unity as a *single entity*, requires a completely different approach than what is demanded by most philosophers, neuroscientists, psychoanalysts, and technicians of computer science and AI. This means that also the approach of trying to establish correlations between the world of neurology and psychoanalysis cannot be used either. Correlations cannot be the building blocks of a model. They can only be tools to get to the *functional* modeling of the Ψ-organ. In principle, the basis must include: *a clear axiomatic, the laws of physics and engineering, and, of course, experiments with the developed functional model.* Without this basis, the idea of bridging the gap between the nervous system and the psychic apparatus remains only a thought experiment, which cannot and must not satisfy an engineer.

Neurology and psychoanalysis explain human behavior by means of different functions, on the one hand the neurological on the other hand the psychological functions, which bring about this behavior (Luria 2001, p. 26; Damasio 1994, p. 14), which generate an individual human behavior. This is the reason why in the previous chapters it was first necessary to elaborate the difference between *functional* and *behavioral descriptions*, and to make the distinction between *complex* and *non-complex* processes. By making these (axiomatic) distinctions, it becomes possible to explain why consciousness arose evolutionarily. These distinctions are fundamental to illuminate the problem from the point of view of technical information theory. In contrast to the previous explanations, however, it must be clear that the hypotheses concerning consciousness have not yet been validated by experiments. Also, our simulation results in SiMA have not progressed far enough, but I have to blame this on the tremendous effort that is necessary to accomplish this and the resistance of my technical colleagues. I will explain this in more detail later.

Let us return to the definitions of consciousness formulated by Antonio R. Damasio and Mark Solms. SiMA is essentially based on them. Antonio R. Damasio formulates in Damasio (2010, p. 157): "... consciousness ... (is) *a state of mind in which there is knowledge of one's own existence and of the existence of surroundings.*". If you follow the argument of Mark Solms: ". . . Sigmund Freud claimed, on the basis of his clinical observations, that consciousness is only a (variable and superficial) *property* of our mind ..." According to Mark Solms (Solms and Turnbull 2002, p. 84), consciousness can only be a property of *information*—and not of functions—because, according to the scientific theory of information, only *functions* (physical and mental) exist in intelligent objects, in which the *information* stored and transmitted in them causes the object itself to act and react. But what kind of information is this? What information has the property of *being conscious? Which have the property of being preconscious* and *unconscious?* And how can we imagine these properties?

Mark Solms first differentiates between a background state of consciousness and consciousness itself. "So, the background "state" of consciousness *does* mean or represent something. It represents "you"—the most basic *embodiment* of your "self"" (Solms and Turnbull 2002, p. 114–115). So, one basis for consciousness is *feeling*, the *valuation of information*. But it is not just a valuation of something, it informs me how I feel *in relation* to objects of the *external world*. So, my mental self-image (the self) is set in relation to objects of the external world, the physical world. Without consciousness, I have no relationship to anything. *Consciousness creates the connection between my psychic apparatus, my self, my inner world to the external world, the physical world.*

At this point I must add something that is essential to me. I have long resisted the concept of *consciousness*. With the statements of Mark Solms it became clear to me. The terms unconscious, conscious, consciousness imply psychic objects, that is, in our sense, functions that do something with information. But as Mark Solms and Antonio R. Damasio make clear, they are *properties* of *information*. This leads to an axiomatic contradiction. It has to be resolved, even though it has existed for a long time and has been burned into the thinking of psychoanalysts. It has also long been believed that the earth is the center of our universe. And it was believed that the Bible was a book from God our Father, which Baruch de Spinoza was ultimately able to refute well, and for which he was ostracized by all three major European religions (Auerbach 2017). I hope it will not end so badly with me when I formulate: "The conscious", "the preconscious" and "the unconscious" do not exist, only *conscious*, *preconscious* and *unconscious* information. Consciousness as an object (function) is also excluded for axiomatic reasons.[12] *Consciousness as a state*, however, can be accepted. But let me continue my thought from above.

My considerations are also supported by Antonio R. Damasio (1999, pp. 24–25): "Consciousness {I, the author, interpret the term *consciousness* in the sense of the *state* of conscious information} generates the knowledge that images exist within the individual who forms them, it places images in the organism's perspective by referring those images to an integrated representation of the organism, and, in so doing, allows the manipulation of the images to the organism's advantage. Consciousness, when it appears in evolution, announces the dawn of individual forethought.". And he goes on (Damasio 1999, p. 127): "*Consciousness* is the umbrella term for the mental phenomena that permit the strange confection of you as observer or knower of the things observed, of you as owner of thoughts formed in your perspective, of you as potential agent on the scene". In other words, *conscious information helps animals and humans to valuate and assess the complex external world, i.e. their own organs as well as objects outside their own body and all their behaviors and states, subjectively and to some extent even objectively.* In order to be able to concentrate on one's own important goals and not to be overwhelmed and paralyzed by a flood of information, the unimportant information is filtered

---

[12] Comment of Volker Hartmann Cardelle: I do not concur with this claim. I consider the question of how consciousness is to be modelled in SiMA an open one and do not agree with Mark Solms' interpretations of Sigmund Freud's view. As it was written above, consciousness has not yet been properly developed in the SiMA-project. Therefore, there is no experimental evidence in favor of Mark Solms' or Antonio R. Damasio's views nor are there any results that contradict the thesis that consciousness is a function within the psychic apparatus, which is, as I understand him, Sigmund Freud's view—a view I personally defend. Experiments will tell us which view we have to abandon.

according to one's own assessments, i.e. subjectively. This is the crucial task of our conscious thoughts and ideas. A person can look into the future with planning because they can subjectively valuate the past and learn to reflect on it and thus learn from it. They learn to valuate history, their actions and the actions of others subjectively and to apply objective criteria. This makes them superior to all other living beings (Ridley 2010, pp. 43–44). It is not for nothing that humans eliminated many other creatures that came too close to him in terms of intelligence, even if they were monstrously poisonous or strong. They were and are, in the long run, inferior to Them in their capacity for consciousness. It is not their muscles, not their speed, that make humans superior to all other living beings. Lions and whales have stronger muscles, and cheetahs and bats are faster. No, it is the ability to be conscious. I became deeply *aware* of this when I read a book by two German geologists. Not wanting to get involved in a war, they fled to the Namib Desert and stayed there for more than 2 years. During this time they had to deal directly with nature in order to survive, and in this sense they tried to deal with it intensively, both scientifically and philosophically. For those who would like to explore the topic of *intelligence* and *consciousness* in relation to animals, I recommend the book by Henno Martin (2009), one of the two geologists. He manages to fascinate his readers with his vivid way of writing.

To attain the state of consciousness is thus the highest ability of an intelligent being, regardless of how we will define intelligence. As Antonio R. Damasio (2010, pp. 13–14) puts it: "It is arguable that cultures and civilizations would not have come to pass in the absence of consciousness, thus making consciousness a notable development in biological evolution.". There is no doubt that industry, the military, and indeed every organization in our society will sooner or later have to deal with the issue of artificial consciousness if automation and AI are to approach human capabilities. But before I explore the exciting aspect of *consciousness* and *machines* in the next chapter, here are some important scientific findings that are necessary to understand SiMA.

Antonio R. Damasio, as a neuroscientist, does not leave the description as it is, but differentiates between core consciousness and extended consciousness. Core consciousness (Damasio 1999, p. 125) "... *is the very evidence, the unvarnished sense of our individual organism in the act of knowing* ...". Thus, it is a state of one or more functions that must be based on a *self* that conveys the "here and now" to the being (Damasio 1999, p. 16). *Therefore, corresponding information must be stored in the self.* With this, the being is not yet able to look into the future and consider its history. But *it can relate itself to the outside world.* These psychic processes can only include events and actions that are happening in the present. To imagine oneself in another place is not possible in the state of *core consciousness.* Therefore, short-term memory plays a central role in this state. It is also crucial for me that there is no conceivable relation to language (word representatives) (Damasio 1999, p. 109), a statement that will be clarified in Chap. 6.

The extent of their *extended consciousness* distinguishes humans from all other mammals (although there is probably not a hard cut, but a gradual transition *regarding the structure of the functions* of the psychic apparatus). Antonio

R. Damasio describes *extended consciousness* as "the complex kind of consciousness" that has many "levels" and "grades" (Damasio 1999, p. 29). This state allows human beings to reflect on their past, to think about the future, to imagine distant places. And most importantly, in the state of consciousness, they can share their thoughts with other people through sophisticated languages.

In this sense, Antonio R. Damasio (1999, p. 17) also differentiates between the *core self* and the *autobiographical self*. The core self is "... a transient entity, ceaselessly re-created for each and every object with which the brain interacts." Yet, consider that Antonio R. Damasio understands by object a something perceived outside the psychic apparatus, that is, a body. In contrast, the psychic *functions* in the state of extended *consciousness* access information stored in the past of the *autobiographical self*. All this information is valuated by quotas of affect and emotions,[13] to which new insights are constantly added. Hence, my idea of myself changes depending on my momentary emotional situation, the moment of my perception, my momentary association of my history, and much more, and it is subject to a continuous process. The information about the whole *self* cannot be grasped in one moment, especially not the *self* from the past. The *self* will always be past. *The self that I perceive at this moment is different from the self of yesterday or the self of tomorrow.* It is subjective and cannot be perceived objectively by myself.

But does this not contradict what was said at the beginning, that one can work out the subjective in a scientific and objective way? Not at all! All neurological and mental *functions* of the Ψ-organ[14] (I am not talking about their patterns) have been scientifically worked out by experts through behavioral observations. These functions as a whole ultimately form the model of the Ψ-organ presented in Chap. 6. If this functional model is integrated into behavioral experiments specified by the experts and simulated on the basis of the Ψ-organ functions that generate their behavior, the simulation results will show whether the experts have correctly worked out the overall functional model or not. If the expected behavior is confirmed by the experiment, one can very well observe and objectively examine the subjective ideas within the model of the Ψ-organ.

And there is something else that follows from this statement that is very important to me. If the *self* of a human being is a continuous process, how can you simulate a concrete human being? The answer must be: You cannot! Not because of the complex structure of the Ψ-organ (we can ignore the rest of the body here), but simply because of the data that have been continuously accumulated in the mental part alone in the course of the history of the individual person, and will continue to be accumulated. How should one determine them? Therefore, one can only simulate the functional model of the Ψ-organ to test theories about behavioral phenomena.

---

[13]This aspect of the stored information, which is always stored in a valuated way—there is no objective storage—will be discussed in more detail in Chap. 6.

[14]With regard to a neurological function, I am thinking, for example, of Broca's area (a function of motor production of language) (Hasler 2012, p. 54), and with regard to a psychoanalytic function, e.g. of the function of defense (List 2009, p. 91).

But I can well imagine Ψ-organs of robots, in which such structures (i.e. functional models) are integrated, and where the states are continuously read out in order to check whether they produce the expected behavior at each time or not. However, I will leave this topic for now and come back to it later.

There is a point that is not insignificant for me, which I would like to add and which has to do with the understanding of consciousness with regard to the modeling of the psychic apparatus. It is the insight of Antonio R. Damasio that the extended consciousness builds on the core consciousness and cannot function without it, but conversely the core consciousness can exist without the extended consciousness (Damasio 1999, p. 18). Consequently, the states of *core consciousness* and *extended consciousness* are based on different, hierarchically arranged *functional units* (entities), which will be the topic of the modeling of the psychic apparatus (Chap. 6).

Yoram Yovell, in his elaborations, presents other models that complement those mentioned above and that can be used to understand certain aspects of the Ψ-organ even better. For example, he Yovell (2008, pp. 121ff) introduces us to George MacLean's three-layer model. This model states that the oldest part is the so-called reptilian brain, which consists of the brain stem, the hypothalamus and the basal ganglia.[15] Most of the knowledge in this area is not learned but genetically determined, which makes sense because the reptilian brain is responsible for the basic physiological functions of the body, such as breathing, controlling blood pressure, heart rhythm, hunger and thirst, sleep cycles, and so on.

Hierarchically superior to the reptilian brain is an evolutionarily more recent part of the brain, which he calls the mammalian brain. It is the mental basis of all mammals and contains the limbic system (hippocampus, amygdala, etc.), which performs functional tasks such as the love between a mother and her offspring, but also the formation of fear. The limbic system is responsible for the creation of subjectivity, which also means that the information processed here is valuated. Which valuation variable is responsible for this is left open here. What is important for our discussion, however, is the task of the valuation mechanism. As I explained above, it consists of judging which information is currently crucial, important, and less important. Consequently, the mammalian brain directly influences the reptilian brain, and is therefore functionally (hierarchically) superior to it.

Yoram Yovell calls the evolutionary youngest part of the brain the new brain, formed by the neocortex. He explains that all mammals have it and in some it has developed into the largest part of the brain, that is the case with dolphins, certain whales or monkeys, and of course humans. In humans, the neocortex makes up the astonishing size of about 85% of our total brain volume (Yovell 2008, p. 129). It may now become a bit more understandable what enormous storage capacity consciousness requires if we have the additional information that the vast majority of our information in the psychic apparatus is not conscious, but unconscious. According to

---

[15] Correcting the term brain to Ψ-organ does not make much sense here, because firstly, brain researchers speak of the brain by such names, and secondly, these elaborations refer mainly to neurological areas (functional units).

Solms (Solms and Turnbull 2002, p. 84), based on measurements by Bargh and Chartrand, the conscious portion is only 1%. However, such numerical values depend on different definitions (axiomatics) and measurements, which are certainly debatable.

I think it can hardly be stated more clearly. In order to be able to process conscious information, i.e. to reach a state of consciousness, which requires the ability to *feel* (Yovell 2008, p. 125), in which objects of any kind are consciously valuated, an enormously high number of neurons is required, which is simply not found, for example, in reptiles or animals with similar brain structures.

The last point I want to make about consciousness is basically a trivial one, which will be immediately obvious to computer engineers, but which reinforces the last point. The following facts may help to understand some aspects more easily.

If we compare the neuronal characteristics of different animal species (Wiest 2009, p. 5, 34), several aspects stand out: the number of neurons and their synapses, and the structures of the different brains. Let us start with the number of neurons and synapses. How did the first neurons evolve? It can be assumed that receptors gave rise to communication transport lines, today's neurons, whose number increased enormously after Charles Darwin's theory of evolution due to their enormous advantages for a living body (Watanabe et al. 2014). Today, one of the most scientifically known creatures with only a few neurons is the sea snail Aplysia Californica (Damasio 2010, p. 32; Kandel 2006, pp. 146–147). Unfortunately, its neuronal structures are already so complex that neurobiologists consider it more useful to simulate only simple neuronal subunits for the time being (Kandel 2008, pp. 17–18) (see also the example of the neuronal circuitry of a grasshopper (Bose 2010)). This bottom-up method helps to clarify basic questions of function, but one should not hope to be able to explore the human Ψ-organ brick by brick in the future.[16] Humans have 10^{12} neurons and 1000 times more synapses (Damasio 1999, p. 396). Because of these orders of magnitude, the bottom-up method will fail for hundreds of years to come. I repeat myself: Only the top-down method will help. This does not mean that the bottom-up method is wrong. It means that the bottom-up method can be used to clarify fundamental questions, such as the scientific question of how the wings of a grasshopper beat. In no case does the bottom-up method help to develop comprehensive, functional models as a whole. This is a tenet of computer engineering. And the Ψ-organ must be developed as a comprehensive functional model.

Ants have about 10,000 neurons[17] (Kriesel 2005). Their intelligence is therefore much higher than that of the snail mentioned above. And if you look at their social behavior, you have to be deeply impressed. But is it really social behavior in the

---

[16]The goal of the Blue Brain project, which I mentioned in Chap. 2, is to simulate the Ψ-organ. But how can you simulate it without knowing the structures? Do they really believe that we can build it up module by module—following the bottom-up method—in order to arrive at a functional overall system?

[17]The information about this varies a lot. In SeiSmart (2019), the number given is 250.00. The sense of my statement stands irrespective of the exact number.

narrow sense of the word? Take human *social behavior*: We understand it as reflective behavior. In order to act in a reflected way one must have an idea of oneself, ergo one needs a *self*. The *self*, as I indicated earlier, is based on an enormous accumulation of patterns (patterns of what is heard, of actions, of what is touched, of images, of smells, etc.). This implies that all this information must be stored. This, in turn, implies that the human Ψ-organ must provide an enormously large memory. Or, to put it more casually: The human Ψ-organ is first and foremost a large associative reservoir in which symbols of patterns are stored, manipulated, and communicated. One can say: The number of neurons of the Californian sea snail mentioned above or of the ant, the bee and all other animals with relatively few tens or hundreds of thousands of neurons cannot provide such a storage capacity. Consciousness, as defined by Antonio R. Damasio and Mark Solms, requires an enormous number of neurons in order to permanently store many of the patterns that flow into our organ over the course of a lifetime.

With this in mind, I often explain the following—perhaps somewhat odd appearing—fact, which makes the crucial point understandable: If you cut a wasp in half, it cannot *feel* pain as we do. *Feeling* requires consciousness. Consciousness presupposes a *self*. But the small number of neurons in a wasp is not sufficient to develop a *self*. One can assume, however, that the wasp also has pain sensors that report the separation of the body to the neuron nodes, which react accordingly (like an automaton?).

In contrast to the previous chapters, this chapter deals with theories which's hypotheses have not yet been (sufficiently) substantiated by (scientific) experiments. The description of these hypotheses and the necessary explanations were and are therefore particularly difficult for me. For this reason, I will try to briefly summarize the decisive points that are especially important for the *functional modeling* of the Ψ-organ.

I am deliberately talking about *functional modeling*, which has always been the goal of SiMA. On the other hand, this does not mean that everything else is not needed. On the contrary, behavioral methods are very much needed for clinical work. But I do not want to say anything about that, since I lack the necessary expertise; that is the job of clinicians. My realization is that consciousness is gained by a valuation (feeling) when the inner *self* establishes a relationship to an object of the external world (in the psychoanalytic sense) and at the same time (in the Freudian sense) assigns a word to it in an abstract way. I can live with this hypothesis because simulated models can be used to conduct experiments by which different theories can be tested. This is the purpose of SiMA. SiMA should help to better understand the ideas and the resulting models of the composition of the Ψ-organ. Of course, these simulation models can also be implemented in robots and will certainly help in the fields of automation and AI.

4.4   Intelligent Machines of Today and Machines with Consciousness of Tomorrow

As an electrical engineering student, I tried to generate speech electronically and build voice-activated controls to turn our lights on and off. This was around 1970, long before electronic components were commercially available. I failed miserably. When I read Sigmund Freud's book (newer German edition: Freud 2001) "Zur Auffassung der Aphasien", I realized that I still lacked quite a bit of knowledge to be able to devote myself seriously to this task. Sigmund Freud states in this critical study that language has to do with consciousness. Language is not possible without consciousness and vice versa. Today I know that this statement can easily be misunderstood. We must distinguish between *word recognition* and *recognition of word meaning*. Today, electronics are so sophisticated that they can recognize words with a certain statistical probability. However, electronics are far from being able to understand the meaning of words (see the *symbol grounding problem* already alluded to in Sect. 3.3). According to Sigmund Freud, this would mean that electronics have to have consciousness. I, on the other hand, just wanted my machine to recognize words. Today, some 40 years later, this works relatively well. But why should machines understand the meaning of words? And is this not problematic? To explain this, I have to go back. The historical origin is to be found in mankind's generally growing natural scientific knowledge, but also in the development of specific knowledge in the fields of automation and, in particular, engineering. I would like to deal with ethical questions not before the last chapter. Then I have explained the model of the Ψ-organ—the SiMA model, which can actually be simulated and could be realized in robots—and worked out the necessary preconditions to be able to go into far-reaching considerations.

The idea that tools enable humans to simplify many things and thus relieve the burden on themselves is an old one. Humans used their first tools as early as the Stone Age. Certain animals also have such abilities. The Ψ-organ thus allows abstractions at a certain developmental stage of evolution in order to facilitate special tasks. One can perhaps define this developmental stage of humans in the Stone Age as an epoch in which humans had a modest understanding of physics. From the point of view of engineering, humans increasingly learned to abstract for this purpose.

Humans invented the wheel more than 7000 years ago, and dams and many other useful things about 5000 years ago. The Babylonians also gained the first mathematical knowledge about 5000 years ago. But the basic, abstract understanding of mathematical and physical laws that led to the first great cultural achievements was built up only gradually over the centuries. It was not until the fifteenth century that the physical relationships were so well understood that they ultimately helped *automation*[18] to have a breakthrough on a broad economic basis. This was also the time of Leonardo da Vinci, whose inventions made a deep impression on me as a

---

[18]The term *automation* can be traced back to the Greek inventor Heron of Alexandria (first century AD), who invented a door control system and described it in *Automata*.

teenager. Just think of his investigations into hydromechanics (Bedini et al. 2005, p. 196), which he based on the knowledge of the Greek explorer Heron of Alexandria, or the design of the first ball bearing (Bedini et al. 2005, p. 286).

We can consider this second epoch of mankind as the time when man was able to acquire a basic understanding of physics and mathematics. But it was the third epoch that led to the *Industrial Revolution*, when it was understood how to use large amounts of energy to support human labor. The very term *Industrial Revolution*,[19] however, also reminds us that every development has its downside, which humans must learn to deal with. The story of the Luddites cannot leave anyone untouched.

Since the twentieth century, we are now experiencing the fourth epoch that has changed and will continue to change the world. It is the epoch in which, for the first time, we humans are able to

transmit, store, and process large amounts of information

through machines.

The decisive first developments of this fourth epoch—known today as the digital world[20]—were made by Alan M. Turing (1936: the theory of the Turing machine), Konrad Zuse (1941: the inventor of the first electrical, program-controlled computer, the Z3) and Claude E. Shannon ((1948), the development known today as *classical information theory*). These developments made it possible to develop these three information processes scientifically and experimentally. Further developments in the field of technical information theory followed. For the SiMA project, the theory of Georg H. Mealy (1955) is crucial. It allows the calculation of electronic circuits according to the top-down design method ...

> ... by strictly separating the physical description from the information-technical description—what I call a strongly abstracted intellectual game—and at the same time coupling both description models via defined interfaces.

For the first time, it was possible to theoretically design and optimize *any kind of digital circuit*. From today's point of view, the "tinkering" of electronic circuits ended with Georg H. Mealy, which I will discuss in more detail in the next chapter. This theory is one of the main pillars of the SiMA project.

Technical information theory was extended by the development of WANs (Wide Area Networks, starting around 1955), LANs (Local Area Networks, starting around 1980), and finally by fieldbus technology,[21] which can be dated to around 1980 (Dietrich et al. 1997). Especially the networking of systems/processes required new information-theoretical models of how intelligent (technical) entities communicate

---

[19]The term revolution, as in contrast to the term evolution, implies the destruction of something, otherwise this epoch would have been called "industrial evolution".

[20]I do not like the term: *digital world*. The crucial point is the process of information processing, understood independently of the hardware. Digital technology—i.e. digital hardware—has only made today's information theory possible, both technically and economically. On the one hand, the term *digital* tempts us to focus solely on the hardware and, above all—and this is very important to me for the purposes of this book—it prevents us from seeing engineering as independent of the hardware.

[21]I deliberately do not go into physical or software developments, as these are of secondary importance for the SiMA project.

with each other, i.e. how these processes can be *described* scientifically. I was working in a large telecommunications company at the time, and at some point I just did not understand the world anymore. Why did technical communication need more than the mathematical laws of communications engineering? The computer scientists and mathematicians who were working on these new model concepts were not really taken seriously by us electrical engineers. Almost the entire communications engineering world felt this way. However, they were opposed by computer engineers, computer scientists, and mathematicians, i.e. scientists who were certainly not experts in the field of communications engineering. To us, they were the "greenhorns". But by 1969, these "greenhorns" (scientists) had developed communication models, the forerunners of the Internet (Arpanet 1969), and used them to send information over the telephone network.

Beginning in 1977, these (I will continue to respectfully call them) "greenhorns" developed a communication model that was to be made available as a "recipe book" to all developers of computer communications. They wanted to have it standardized by the CCITT (Comité Consultatif International Téléphonique et Télégraphique, now ITU, International Telecommunication Union), an international organization responsible for communications technology. However, this organization rejected the ideas of the "greenhorns", so they were forced to switch to ISO (International Organization for Standardization). ISO has standardized almost everything internationally, including these model ideas of how systems/processes can communicate with each other. From today's point of view, this was a more than embarrassing attitude on the part of the CCITT decision-makers at the time, who were mainly electrical engineers and especially communications engineers.

But one should not be surprised. Established experts often reject completely new approaches. History has much to offer in this regard. An internal Western Union memo from 1876 is said to have stated: "The telephone has too many weaknesses to be seriously considered for communication". Or, as we know, Thomas Watson, head of IBM, predicted in 1943 that there would be no more than five computers on the planet. Today, my cell phone already contains three computers. And I myself not only misjudged the ISO communication model, which I later taught at universities for more than 30 years, but I made an even more serious misjudgment. As head of the department of communication devices, I was once asked what it would cost the company to develop a portable telephone. I asked our sales department how many units they could sell in the next few years and what such a device might cost. I asked my developers to calculate how much we would have to invest in development. And the result was obvious to me: The portable telephone (now called a mobile phone) would not have a chance in the near future. And how wrong I was. It took only a few years for even small companies to successfully bring their bold developments to market—unlike my company.

And what is the consequence of the wrong decision regarding the communication model? The ISO-standardized communication model, the theoretical basis of almost all technical communication models (see also Sects. 2.1 and 5.3), is now officially called the ISO/OSI model instead of CCITT/xxx model (Walke 1987).

In my lectures on engineering in the 1980s and 1990s, in which I also referred to the ISO/OSI model, which had long since been standardized and used in practice, I not only repeatedly fell on deaf ears, but often also aroused active resistance to the new and to change. I was still confronted with questions such as: Why does telecommunication need communication models, are mathematical representations not enough? Why do people need intelligent machines? What is the advantage of connecting them in a network? When I spoke about the fact—which did not happen often, because in industry one cannot afford to go out on a limb—that in the long run we will not be able to avoid giving machines even consciousness, I was generally considered to be crazy.[22] Today, we know that we cannot do without AI in the Internet and in automation. We also know that, for market reasons alone, it is a necessity that our smartphones are able to connect to all things (Internet of Things) as far as possible. We know that the trend is for large combine harvesters to do their work as autonomously as possible, for drones to find their destination on their own, for people to use Alexa or Siri to switch on as much of their household as possible, for robots to enter burning houses instead of firefighters, and for robots to do their work in a supportive way not only in industry but also in hospitals and nursing homes. But what kind of intelligence is needed to control, monitor and coordinate all these complex processes? Do we really need automated recognition of text and speech content, i.e. consciousness in machines?

To do this, we first need to analyze the different principles, which I think is easiest to do by looking at history and comparing machines to biological, intelligent beings.

Very early on, electrical lines were used as signal and control lines (Fig. 4.5a). Many different sensors (s) were possible. For example, switches, temperature or pressure sensors were used. Relays, lights, or motors were considered actuators (a). Then the possibility of wireless data transmission was recognized, until finally the electronic possibilities were discovered, which meant that data could not only be *transmitted*, but also *amplified*, *stored* and even *processed* (Fig. 4.5b). This also means that information,[23] even from different sensors, can lead to new information and be passed on to different actuators (Fig. 4.5c). Components such as simple electronic units (e in Fig. 4.5d) or central electronic units (CU in Fig. 4.5d) form the basis of various networks.[24]

The distinction between electronic units and central electronic units is given here only as an example to show that the distribution of tasks in such units can be very different. In practice, however, the naming is usually application-specific, which allows a better understanding of the respective information systems. Instead of pure hardware electronic circuits, however, programmable computer modules are almost

---

[22] I have to admit that at that time I did not really know what I have to imagine by consciousness and when this would become reality. But I knew that we could not simply put off the problem.

[23] The relationship between information and data: By assigning meaning to data, it becomes information. Conversely, information requires physical carriers in order to be able communicate, store, and process the data.

[24] An example of such an electronic unit would be programmable logic controllers (PLCs).

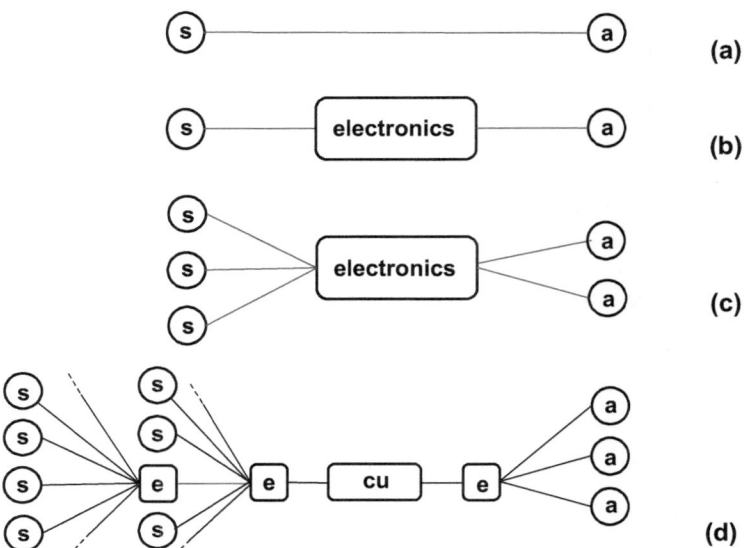

Fig. 4.5 Information transmission, processing, and storage over the course of the development of engineering (s: sensor, a: actuator, e: electronics, CU: central unit). (**a**) Transmission of information over a single line, (**b**) integration of electronics, (**c**) multipoint interconnection, (**d**) creation of extensive networks

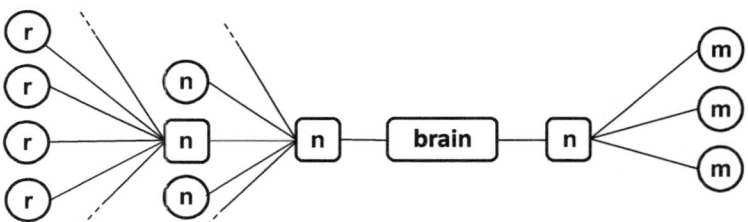

Fig. 4.6 Information transfer, processing, and storage in the biological nervous system (r: receptor, n: neuron, m: muscle or gland)

exclusively used today, which has already been justified in detail in Loy et al. (2001) for the automation of buildings and industrial plants.

If we compare electronic networks as shown in Fig. 4.5 d with networks of human neurons and visualize them as we are used to in technology, we arrive at pictures like Fig. 4.6. In neurology, the sensors correspond to receptors (r), actuators to muscles (m) and glands, and the central unit corresponds to the brain. This representation shows in principle not only the same as Fig. 4.5d, but also that of Fig. 3.2. The developers of automation thus arrived at structures of engineering as we know them from biology and especially from neurology. The *only* task of such networks of technology and neuron networks is the *transport*, *processing* and *storage* of information.

In this sense, a computer in technical networks is a lot of electronics in a very small space, and thus a powerful information machine that is programmable, i.e. can be used in a flexible way. However, the principle of programming must not be misunderstood. In technical products, a desired process sequence can be flexibly determined and the electronics can be flexibly adapted to the respective desired process[25]—a decisive reason for the success in the history of the computer. In contrast, the human Ψ-organ cannot be programmed by humans themselves like an artificial computer. The so-called "self-programming" of the Ψ-organ (when learning to drive a bicycle or a car, or for example after a stroke) I must exclude from the discussion. So far, I have no knowledge that I could use without further effort in the sense of SiMA. But leaving aside the question of programming, I have to state that from the point of view of information engineering and the regulation of a process, one can very well compare the artificial computer with the Ψ-organ of a human being. Both are information systems. Computers in washing machines, rockets, cars, radios and mobile phones have the only task to control and regulate these systems, i.e. to process information in them. The Ψ-organ has the task of ensuring that the *homeostasis* process of a human being remains reasonably regulated, i.e. can be kept in balance even in the long term. All this takes place in the form of information processing in neurons. In this process, the Ψ-organ (which in this sense should be considered as a concentrated set of neurons or as an information storage) processes the incoming information with the stored information and makes decisions about it in order to activate muscles and glands. It leads the human being to actions (see again Fig. 4.5d).

I cannot emphasize enough that the term *computer* is defined and must be understood in an abstract (i.e. mathematical) way, as an *information-processing* machine. We must therefore first detach ourselves from the underlying structures, topologies, hardware, etc. and how information processing is carried out in detail. In SiMA, I want to put the focus on information processing, because that is the task of an Ψ-organ (a biological computer) as well as that of an artificial computer.

I think the first AI scientists must have thought in such a simple way. The predictions they made at that time, as I mentioned at the beginning of the first chapter, could only be too optimistic. They correctly understood how one can understand what the main task of the Ψ-organ is in *principle* and compared it with industrial processes. They relied on models, of which most are still valid today. Norbert Wiener was one of the outstanding American pioneers to be mentioned in this context (Wiener 1970; Frank 1970, p. 14, 15; Hassenstein 1973, p. 56). But they gravely misjudged and underestimated the enormous difference between the task set for industrial processes at that time and the task set for a Ψ-organ. Above all, the focus was on hardware, which is why robotics, with all its mechanical challenges, was and is often disproportionately emphasized. But one has to realize: *The decisive*

---

[25] An efficiently applicable "self-programming" of the processor on the basis of "cognition" is far from being conceivable. For this purpose, appropriate models are missing, which will be discussed in the last chapter.

unit that distinguishes humans from animals is not their body, but their superior control, their psychic apparatus. The muscles, the locomotor apparatus, the possibilities of energy production, the physiological principles of maintenance cannot be overestimated, but the decisive organ of humans is their Ψ-organ and in it the *psychic apparatus*, not their muscles. The psychic apparatus makes them the supreme living being. I repeat myself: Lions, whales or cheetahs, to name just a few animals, are certainly physically superior to humans in many ways, but not in terms of the power of their information processing in the Ψ-organ. And the psychic apparatus embedded in it is far from being sufficiently explored. At the time when the new scientific field of *AI* was founded, some knowledge of psychic phenomena was already available, but a holistic and comprehensive functional model of the psychic apparatus was unknown in natural science—except for Sigmund Freud's model. But his *comprehensive*[26] *functional model* was not recognized[27] by medicine and even less by the natural sciences such as physics or chemistry.

However, this resistance to the psychoanalytical model must be countered by the following basic consideration. From the point of view of the present state of knowledge of computer technology and computer science, a *top-down approach* must be the prioritized approach. This means that if one wants to understand the human being and especially their Ψ-organ, one must start with the psychic apparatus when modeling the Ψ-organ, and *to this day the only science that has worked out a functional, holistic (comprehensive) model is precisely psychoanalysis.* Consider the preface of the book (Freud 2001, pp. 7–31) by Paul Vogel (professor of neurology at Heidelberg University) and Wolfgang Leuschner (former deputy director of the Sigmund Freud Institute in Frankfurt).

Thus, although the founders of AI had the right vision of where the journey had to go, they were—and I would argue that many scientists are still today—too "rooted" in the descriptive tools of physics and chemistry to be able to break away from their traditional way of thinking and study the information theories of Claude E. Shannon or Georg H. Mealy without prejudice and adopt them as new tools of natural scientific thinking. And I cannot emphasize enough that the ways of thinking in physics or chemistry and the ways of thinking in information theory are very different, even though mathematics is the basis of both. But mathematics cannot describe everything, even though that was the dream of Baruch de Spinoza and many others, and often still is.

---

[26] I would like to point out again that the term *comprehensive* here does not mean that it is *complete* in the sense of a detailed functional description, but that it describes all the crucial functional elements of the psychic apparatus. The term comprehensive is therefore to be understood in the sense of holistic.

[27] Even in the general literature of the disciplines of psychology (Solso 2005; Baars 1989) or medicine (Churchland and Sejnowski 1997; Eccles 1975; Popper and Eccles 1977; Singer 2006), but also in cybernetics and AI (Bammé et al. 1983; Breazeal 2002; Tan and Nijholt 2010; Erismann 1968; Forth and Schewitzer 1976; Klaus 1969; Lunze 1995; Schank and Childers 1986, Weizenbaum 1976), psychoanalytic models are given little space. For the most part, psychoanalysis is simply ignored (Palensky 2008; Schülein 1999, p. 37).

How should AI be viewed from today's perspective compared to conventional computer technology? The answer to this question makes it easier to understand why certain machines need to be conscious.

I would like to exclude the PC (personal computer) here, as it does not generally regulate processes autonomously. Most computers today are integrated into industrial controls, appliances, equipment, cell phones, automobiles, buildings, etc., all of which perform specific process tasks in technical processes. I want to talk about these units here. They receive information through data inputs and provide output information through algorithms, that is, through their data processing. For this purpose, we engineers use mathematics to describe the processes, which requires that we can abstract the processes to such an extent that, among other things, the necessary sensory input information is available for the equations used.

I would like to return to Sect. 4.1. If the processes are complex (complex according to my definition), it is no longer as easily possible to develop mathematical algorithms. If one tries it nevertheless, by simplifying the description increasingly, i.e. abstracting more and more, this means that the inaccuracy increases as well. As a result a control system developed in the classical way can no longer function sufficiently well. This is the reason why, in the end, humans still make the decisions in many automated processes, i.e. they intervene in processes as *controllers* and *decision-makers*, so to speak. Let me give you an example.

Autonomous vehicles are a hot topic today. As long as a vehicle runs on rails, one can imagine that the processes that a computer needs to control can be described more or less adequately in a physical way: reducing the speed of the train when it enters a curve, detecting and responding to stop signals, or initiating emergency braking under certain conditions, and so on. But it will be difficult to integrate the right sensors for all conditions, or even to take all conditions into account. This is where AI comes into play. It should be able to cope with situations and processes that cannot be calculated in advance. In other words, AI should help us when we do not have enough information to describe a situation physically or chemically with reasonable accuracy. AI therefore relies heavily on *statistical methods* or *learning algorithms*. These are mathematical methods that generally belong to the scientific field of computer science. They try to reproduce the *behavior* of animals and humans for specific situations.

This is more or less the state of the art today. However, this does not yet answer the question of why machines should have not only this kind of AI, but even an AI with consciousness. Should or must a machine perform mentally highly demanding tasks?

Let us recapitulate. The tasks of the Ψ-organ are manifold. On the one hand, it has to stabilize the homeostasis of the body, but on the other hand, it has to think about how to proceed in the near and distant future. This makes it clear that the Ψ-organ can only be seen in a multifaceted way. The electronics, computers, and AI that computer engineers and computer scientists have developed so far, however powerful they are, must be seen as very modest compared to human mental performance. As I explained in Sect. 3.2, humans do not think in the form of mathematical algorithms—as many technical controls do—but instead—in addition to their apparently[28] logical conclusions—they associate previously experienced ways of solving a task that led to success or non-success. In this process, the so-called "gut feeling" (i.e. unconscious patterns) often plays a decisive role.

With these considerations in mind, we must ask: Where do humans still need to intervene in processes that we would rather leave to machines, which might do them better? I am thinking, for example, of a project that we led (Dietrich et al. 2008): monitoring the Krakow airport in areas accessible to passengers. Is there a suspicious object (suitcase, bag, or something else) that could be dangerous? Are there any sounds that indicate a dangerous situation? Our task was to solve these questions with the help of AI. At that time, I was not satisfied with our solutions because the monitors showed such dangerous objects, but in the end it was still the human who had to decide whether the police should intervene or not.

Or let us take another example: The bus driver in the city or even in the countryside. Sure, we can have fully automated buses today, if the locations can be sufficiently developed and clearly secured. But here in Berlin, for example, we do not expect the very demanding job of a bus driver to be replaced by intelligent, autonomous machines in the next few years. Children suddenly running into the street or poor lighting at night in snow or rain are challenges that require a high degree of mental agility from the driver. And when we think about language, the problematic nature of machines with consciousness becomes even more apparent. If you search in Leo or DeepL other electronic dictionaries or translation programs, you will of course get a lot of English terms for a word like "think". Of course, the programs are getting better and better at finding the desired term in conjunction with other terms in a sentence. But a translator does not work with lists to translate texts. They try to *understand* the content, like a bus driver has to understand the situation on the road, like a security guard has to *understand* the situation in the airport waiting room in order to act accordingly.

What does that mean from a psychoanalytic point of view? For in SiMA we take the psychoanalytic model as a basis. What does *understanding* mean? In the previous chapter, the concept of *consciousness* was explained as best as possible on an axiomatic basis. *Understanding* has to do with consciousness on the one hand, but also with the *feeling* that *valuates* objects, situations, indeed everything of which one gains an idea. What is my relationship to these objects or situations? And behind the feelings, of course, there are other valuation mechanisms, such as *affect quotas* and *emotions* of various kinds, which I will discuss in detail in the next chapters.

I think that makes it clear. The field of AI cannot and will not remain at its current level. It must deal with concepts such as *consciousness* and *feeling* if the technology, it develops, is to take over more tasks from humans. Consciousness requires the highest form of intelligence because it combines two mental principles. On the one hand, when a person has to solve a task, they always associate memories of their

---

[28] I use the term "apparently" on purpose, because one can very well argue whether logical reflection or rather affective, emotional and emotionally valuated associations have the upper hand in a given case.

success or failure in similar tasks, which produces an *emotional* and *feeling valuation*. On the other hand, an adult has learned to think causally and logically. This makes the human being the superior being on our planet. If we want to advance automation, we need to understand how humans work.

And psychoanalysts, like all other natural scientific disciplines, need a simulation tool to clearly verify their findings and continue their research.

SiMA stands for all of this.

Chapter 5
Models and Tools of Computer Engineering for Bridging Neurology and Psychoanalysis

It is crazy! Many in AI believe they can understand intelligence through algorithms. Neuroscientists are desperately searching for classical physiological models and explanations to explain mental functioning (Hasler 2012). Psychologists and psychoanalysts are stuck in their limited world of information. And then some believe that finding correlations between physiological models of neurology and psychoanalytic concepts is the silver bullet. But since the methods of description and therefore the laws of physics and chemistry are different from those of information theory—to which I count psychoanalysis—this cannot work. But everyone sticks to his own field. Everyone believes to be able to find the solution for the bridging with his familiar method of description or declares it to be undiscoverable. An intensive scientific examination of a field other than one's own is rarely carried out. But this is my long experience in standardization committees and with students and scientists of all three fields: One has to deal intensively with the different fields, or let us say, as before, "worlds". You have to understand and apply their methods and laws in experiments, otherwise you will hardly gain access to the other sciences, and the bridge between neurology and the psychic apparatus will remain closed to you. In other words, one has to work very hard to gain access to the respective ways of thinking of psychoanalysis, neurology, and computer engineering/information theory. None of my project collaborators, whether computer technician, neurologist, or psychoanalyst, has managed to do it by reflection and thinking alone. Our experience has always been that only working with experiments helps to think outside the box.

I have already discussed neurology, symbolic information processing, and psychoanalysis. In present chapter I would like to focus on engineering, i.e. information theory from the point of view of computer engineering. For some beginners it seems simple, but when it comes to the practical implementation, one soon encounters difficulties. Therefore, I will give a more detailed description. As mentioned at the beginning, I deliberately leave out the mathematical derivations that I taught for decades as a professor of computer technology. They are not necessary to understand what is relevant. They are only needed for the detailed practical application and realization.

D. Dietrich, V. Hartmann Cardelle, *Simulating the Mind II*,
https://doi.org/10.1007/978-3-031-69530-8_5

But I have to mention one thing in advance. I am certainly not the first to come up with the idea of describing the psychic apparatus in a natural scientific way [see, for example, (Peterfreund and Schwartz 1971) or (Turkle 1988, p. 244)]. Starting from the consideration that a clinical method requires a robust model as a basis, the question of how *to bridge the gap* between *neurological* ideas and the psychic considerations of *psychoanalysis* has been at the forefront of psychoanalysis from the very beginning (Freud 2020b, p. 276). How can one comprehend with a single model how electrical impulses of neurons and thoughts in the secondary process are related? Can there be such a unified model of both different "worlds"? If so, how can it then be e*valuated* on the basis of natural scientific methods so that it can ultimately be *implemented* in robots?

Thus, in the context of the so-called "crisis of metapsychology" (Holt 1973), Emanuel Peterfreund took a consistent, original, but arguably controversial position in the 1970s and 1980s.[1] In the USA at that time a group of psychoanalysts was forming who openly questioned the scientific status of psychoanalytic metapsychology (Holt 1973, 1985; Gill 1976) *Psychoanalytic metapsychology*, especially that of Sigmund Freud, was said to be conceptually ill-defined and contradictory. They said that it used old, metaphorical concepts from biology, physiology, and anatomy that were neither logically consistent nor did they form a closed theoretical system. Emanuel Peterfreund wanted to oppose this view. He did not want to abolish metapsychology, but to put it on a new basis in this sense (Peterfreund and Schwartz 1971, p. 10). He argued that precisely because the clinical practice of psychoanalysis had proven to be valid and useful, there had to be a meaningful theory to accompany it. He argued that from "low-level anthropomorphic concepts" a "high-level" theory had to be developed that would stand up to general scientific standards (Peterfreund and Schwartz 1971, p. 64). Emanuel Peterfreund, working with the computer scientist Jacob T. Schwartz, promoted systems and information theory models that redefine the concepts of *structure, process, function*, and history on which Freudian metapsychology is based. Other principles of Emanuel Peterfreund's reconceptualization of psychoanalytic theory include:

1. The psychic apparatus can be described as a hierarchical system of information processing units, which are subject to the general logic of information theory (Peterfreund and Schwartz 1971, p. 135).
2. The psychic apparatus is therefore something that can be described and grasped with known principles of natural science, which is probably another blow for mankind.[2]

---

[1] That is why I mention Emanuel Peterfreund. I do not know of any other persons who pursued similar goals and are worthy of mention.

[2] See in this regard: (Reppen 1981, p. 159) and (Schülein 1999, pp. 37, 93). Sigmund Freud speaks of three narcissistic blows of mankind: the cosmological, the Darwinian, and the psychoanalytical (Freud 2020c, pp. 27–28).

3. The connection between the *body* and the *psychic apparatus* can be explained on the basis of information theory, which unites both areas monistically (Peterfreund 1980, p. 340).

I would like to claim that the principles formulated by Peterfreund are highly remarkable, but that he was bound to fail.[3] I can substantiate this claim from *today's point of view* with the following arguments:

The information theories known at the time when Emanuel Peterfreund developed his theories, especially the contributions of Norbert Wiener and Claude E. Shannon, are essential for computer technology. However, they represent only a small part of today's information theory and are not sufficient to bridge the gap between neurology and the psyche. Emanuel Peterfreund's colleagues recognized this even then and emphasized that his method was not based on Wiener's/Shannon's information theory (Reppen 1981): "It should also be noted that Emanuel Peterfreund's work is not information theory in the Wiener-Shannon sense but that it bears closer resemblance to cognitive and Piagetian psychology, the work of John Bowlby, and more sophisticated systems theorists and model constructors." And accordingly (Friedmann 1975): "... the Wiener-Shannon theory speaks only of a special measure of quantity of information. It does not deal with differences, with how the information is processed, with the significance, relevance, or meaning of the information, nor with the response of the organism to the information—all topics of prime interest in psychology." In order to solve these problems, theories have to be applied which have been developed in the last 30 years of computer and communication technology and which have to be generalized in order to bridge the gap between the neuronal system and the psychic apparatus. This is what needs to be done in this chapter. These models and tools need to be understood.

There is one fundamental thing to be kept in mind here. As a computer engineer, I had to develop computers for many different fields of applications: for aviation, for household appliance, for the automobile industry, for geological fields, for banks, for power engineering, among others. One thing I had to learn in the process: If you want to develop scientific models that are to be functionally simulatable and ultimately emulatable, no matter for which area, you must not try to bend the theory of the area in question so that it is convenient for us computer engineers. We are not the specialists in aviation, household appliance, automobile industry, etc. I find it hybris, if a computer engineer presumes this task, even if some of them imagine to be able to do it.

> Our only task as computer engineers must be to take the ideas of the experts in the respective field and convert them into a natural scientific model in such a way that this model does not contradict the ideas of the expert.

---

[3] I would like to emphasize that I have the highest respect for persons like Emanuel Peterfreund, who recognize fundamental problems or contradictions in a theory they hold dear and try to work on solutions, even if it seems almost impossible at first.

There is only one exception: If there is a contradiction with natural scientific thinking, a solution must be found *together* with the experts. The suggestions should *always* come from the computer engineers.[4] The experts in the respective fields, in SiMA mainly psychoanalysts and neurologists, must then take on the difficult task of correction, which can only be solved sensibly in several steps as a team together with the computer engineers. This approach has proven itself in SiMA, not the other way around.[5]

Thus, on the one hand, Emanuel Peterfreund's idea of involving the mathematically oriented scientist Jacob T. Schwartz to support him was correct, but on the other hand, the mapping of the psychoanalytic concept into a scientific model should have been done by Jacob T. Schwartz himself with the support of Emanuel Peterfreund.

Emanuel Peterfreund, however, wanted to accomplish two steps at once: a reorientation of psychoanalysis and its mapping into a purely natural scientific model. He therefore took the lead as a psychoanalyst. But this was a fundamentally wrong decision. He had neither the training of a natural scientist nor the experience to conduct natural scientific experiments. For in the development of a natural scientific model (especially of a functional model) the following question must always be in the foreground: How can and must the result be evaluated by experiments? I miss this approach in Emanuel Peterfreund's transcript (Peterfreund and Schwartz 1971), and it is precisely this that is crucial for natural scientists. We must also ask how Emanuel Peterfreund himself reflected his ideas. With today's simulation technology, model results can be tested experimentally very well. Emanuel Peterfreund did not have this possibility. His considerations were still based on the technology of Fortran programs, at that time one of the most powerful programming languages for scientific simulations, but not comparable to today's software languages and tools as used in SiMA.

So, how should we proceed from today's point of view on the basis of Emanuel Peterfreund's demands? This can be formulated quite clearly:

> The theories of psychoanalysis are to be cast into a scientifically oriented, holistic functional model by computer engineers under the guidance of psychoanalysts and to be tested and evaluated by simulation experiments that must be specified by psychoanalysts. Psychoanalysis must be the first to be considered, since it represents the top of the Ψ-organ in terms of top-down design. Neurology describes the bottom layer which can only be developed once the layers above it are at least rudimentally developed.

---

[4]The reverse approach, psychoanalysts proposing solutions first, has proven far more time-consuming.

[5]For example, Thomas Jürgensohn discusses this topic with a focus on human-machine systems (Jürgensohn 2006). He also focuses on the collaboration between psychologists, who mainly use statistical methods to gain insights, and engineers, who are interested in developing devices for humans. So, he is using behavioral models as a basis, which means that he is dealing with a completely different topic than we are in our SiMA project. In SiMA we are developing a functional model. This must not be confused. Indications of interdisciplinary work must therefore always be taken with a grain of salt, otherwise one quickly falls into a trap.

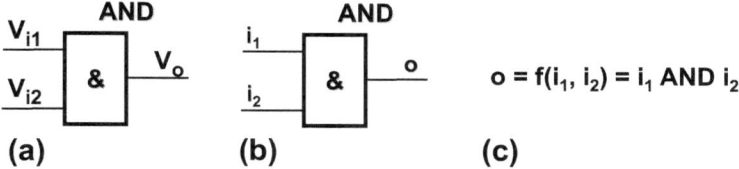

Fig. 5.1 *AND*-function. (**a**) *Callout with voltage values,* (**b**) *formal logical description,* (**c**) *mathematical equation*

> This development process is explained in the following by the theory of George H. Mealy's theory.

Once the model has been established and extensively tested and evaluated, it can serve as a tool for psychoanalysis in the second step of the research to make further considerations that directly and indirectly include the reflections of Emanuel Peterfreund.

In the following chapter, I will first briefly discuss the concept of information theory. Most importantly, how is the term "information" to be understood? What does it mean in the context of natural scientific laws? Which are to be applied in SiMA? Once these questions have been addressed, the theory, that lies at the heart of the SiMA project, must be understood: the Mealy principle. On this basis, I would like to describe the layer model of communication technology, as the SiMA model can be derived from it. In the last subchapter I will introduce the abstraction model, which we also call the model of abstraction levels. This model helps to scrutinize the individual layers of the layer model in different levels of abstraction, i.e. to describe the individual layers in a more or less abstract way.

5.1   Information Theory

These are the elementary questions for understanding the *extended Mealy model*, which is explained in the next central chapter of the book: What do we mean by *information*? How does information relate to the body? What is a physical description, what is an informational description? Answers can be obtained deductively, i.e. starting from mathematical formalisms that are validated, or inductively, by building on what is known and developing the considerations step by step. I want to use the inductive method because I have rarely fared well with the deductive method when trying to give an explanation. I will start with the explanations given in Sects. 2.1 and 3.1.

Let us look again at Fig. 3.15b and take as an example the electronic circuit *AND*, which is to be regarded as a functional unit, i.e. as an example of a functional model. It is shown again in Fig. 5.1. How does this function, this model, behave? Let us assume that the rectangle in the picture, in which an &-sign is written, is a black box. How it is internally functionally structured, i.e. which components (individual functions) it consists of, is up to us engineers. On the other hand, it must be understood that the &-sign means that the function of the module is the logical AND. First of all: What does it mean *physically*? What is the connection from an electronic point of view? For such devices, it has been established that the voltages at the inputs and outputs can only assume two different voltage levels (except for the dynamic transitions, which are not of interest here), hence the name digital technology. For example, let us assume that the two voltage levels to be defined in this circuit are $U = 0$ V (U: voltage, V: volt) and $U = 5$ V. Then the AND function in Fig. 5.1 says if 5 V is applied to both inputs i_1 and i_2, the output o of this black box will also be 5 V. If, on the other hand, one of the inputs is 0 V, the output of this black box will also be 0 V. This corresponds to the international convention of a logical AND function, which was already explained in Fig. 3.11.

What is striking about this explanation? On the one hand, we are talking about the physical quantity voltage, and on the other hand, I assign the symbolic information quantities 0 and 1 to the voltages in the table in Fig. 3.11. In effect, I am translating states of the *physical world* into states of the *information world*. Now let us try to look at this in a more differentiated way.

Figure 5.1b is the same circuit as Fig. 5.1a, but it can be interpreted electronically (= physically) as well as informationally. From the information point of view, the inputs and output i_1, i_2, and o are symbolic values. This means that in Fig. 5.1b you one has to decide whether you want to look at the circuit in the physical sense or in the information-technical sense. If one chooses the physical approach, the laws of physics apply; if you choose the information engineering approach, the laws of information engineering apply. If the laws of physics are applied, for example, the corresponding currents, powers and resistances of the circuit can be calculated. If, on the other hand, the laws of information engineering are applied, the laws of information engineering can be used, namely the equation in Fig. 5.1c: $o = f(i_1, i_2) = i_1$ AND i_2 or the table in Fig. 3.14d. The output function o is a function of the input information i_1 and i_2. These two input information variables are combined by the *AND* function.

Likewise, one could build an electronic circuit based on two different current levels instead of two different voltage levels. In terms of information engineering, this leads to no change, but electronically (= physically) it does.

For the description of physical facts decimal numbers are preferred, in the digital world binary numbers. This means for the circuit in Fig. 5.1 in information engineering terms that to the two voltage values of Fig. 5.1a in the information engineering world of Fig. 5.1b the two logical symbols (*information quantities*) 0 (zero) and 1 (one) are assigned. The symbols represent the *information* about physical states, which we can calculate with via the binary system. This is the basis of all artificial computers.

The corresponding mathematical formalism from the point of view of information engineering is shown in Fig. 5.1c. It states: The output o is a function of the two inputs i_1 and i_2 (Fig. 5.1b). Physical quantities cannot be derived from the information engineering equation. Therefore, it is not possible to deduce whether the information system is an electronic circuit or a non-electronic device such as a water clock.

This leads me directly to a wonderful example: Thousands of years ago, the Egyptians used water clocks to clearly time court proceedings. They filled a vessel with water, which had several vertical openings. The distance between the openings marked a precise time period, which meant that time could be measured fairly accurately. They did not use electricity as a physical medium to measure time, as we computer technicians do today, but water. However, this water clock can be described in both physical and computer terms. However, the computer description of such an information system is quite banal. If the vessel has five holes in certain vertical distances to each other and the water always flows off between two holes in 5 min, one can formulate *physically*: $4 * 5$ min $= 20$ min. *Information-technically* one can say: The information about how much time it takes for the water to drain between two holes is defined as the information quantity a. The information quantity a is the total time for the water to drain. Then the total time for emptying between all holes is obtained: $5 * a = 5\,a$. In order to connect this decimal formalism with that of the *AND*, *OR*, and *NOT* functions, the information quantities must be transferred from the decimal system to the binary formalism, in which only the two numerical values 0 and 1 exist. If the water has run empty up to the first hole, the binary numerical value 1 is reached, after the second hole the binary value 10 and after the fifth hole the binary numerical value 101. The informational formalism of the Egyptian water clock thus corresponds to an electronic binary counter of the numerical representation from 0 to 101, which can be realized with electronic *AND*, *OR* and *NOT* functions.

What is the point of this long explanation? The information functional units (like an electronic circuit or an Egyptian water clock or whatever physical object) and their behavior can be described either by purely *physical* or by *information-technical* methods. *Physical methods* use quantities such as *electric voltages* or *quantities of water*, while *information-technical* methods use a (further) *abstracted* symbolism that is detached from physics, for which numbers of the decimal or binary system are used. Compared to physics, chemistry, etc., the methods of information engineering represent their own, highly[6] abstracted descriptive space. These methods are, so to speak, tools to make connections more understandable and manageable. Of course, mathematics must be used as a further tool for both scientific fields, the physical as well as the information engineering. Once again: it is possible to understand an electronic circuit or a water clock without using the methods of information theory. Theoretically, a computer could also be described in purely physical terms, but only theoretically. In the case of the water-clock, the purely physical description is still easy to understand; in the case of the electronic circuit, it can become "very" confusing. With the artificial computer it becomes much too elaborate and complicated, and for our Ψ-organ much too complex. It is easier to use the tools of

---

[6]I deliberately use the term "highly" here, because information is defined in a way that is detached from matter, and the physical descriptions themselves are already abstract descriptions.

information theory (i.e. its description methods). That is why they have been developed. Computers become calculable with them and thus easier to handle.

With information systems, such as the computer or our Ψ-organ, we can first detach ourselves from the physics and consider the hardware separately. But we must always keep in mind the *connection* of the different languages of description and the methods how to describe and formulate this connection. This is the task of the next Sect. 5.2. In natural sciences, it is not allowed that two such worlds exist unconnectedly (i.e. independently from each other) next at the same time. This was an unsolved problem for Norbert Wiener and Claude E. Shannon. This problem was solved theoretically by Georg H. Mealy.

Thus, information is an abstraction in which physical or chemical methods of description are of no interest. Physics, physiology, etc. only describe the carriers of information. For this reason, information theory can be understood in a very broad sense, because it includes all conceivable information systems. The arc can be stretched from technical information theory (Shannon 1948; Steinbuch and Rupprecht 1967) to the humanities and even to philosophy. In this book, however, I want to limit myself to those parts of information theory that are necessary for the first steps towards a holistic, functional modeling of the Ψ-organ, i.e. the *neurological* part in connection with the *mental* part.

In general, information theory is concerned with how *information is processed, stored, and transmitted*. The central task of computer science is the formal description of information and its processing. While computer engineering—as it has always been my understanding as a professor of this subject—is concerned with the description of hardware and software[7] as well as with the interface (many call it the "bridge") between the physical and the information worlds. Unlike computer scientists, we computer engineers should never lose sight of the hardware, i.e. the carrier of the information, but also of the software (information engineering) that is mounted on it. Figure 5.2 illustrates this distinction.

I would like to summarize in this respect: When I speak of the fields of physics, chemistry, psychoanalysis, neurology, or information theory, I always see them as *specialized fields* with their *laws* and *methods* for understanding these different fields of knowledge. They are thus—more technically formulated—different *forms of description of something*. I differentiate especially between the fields of natural sciences such as *physics, chemistry etc. on the one hand* and *information theory on the other hand*. Today, I see much more behind the term *information theory* than what Norbert Wiener and Claude E. Shannon formulated in their fundamental considerations. Especially with regard to the SiMA project, I also see a necessary, clear, axiomatic definitional space that needs to be worked out for psychoanalysis. I see the scientific methods and findings developed since Norbert Wiener and Claude

---

[7]In the case of the technical computer we speak of hardware and software, where software is understood as the language in which the functions of the software are written. In the Ψ-organ we do not know any tool or language to describe the functions of the information system. Therefore, here we directly name the function in which the information is *processed, transmitted and stored*, and that is the *psychic apparatus*.

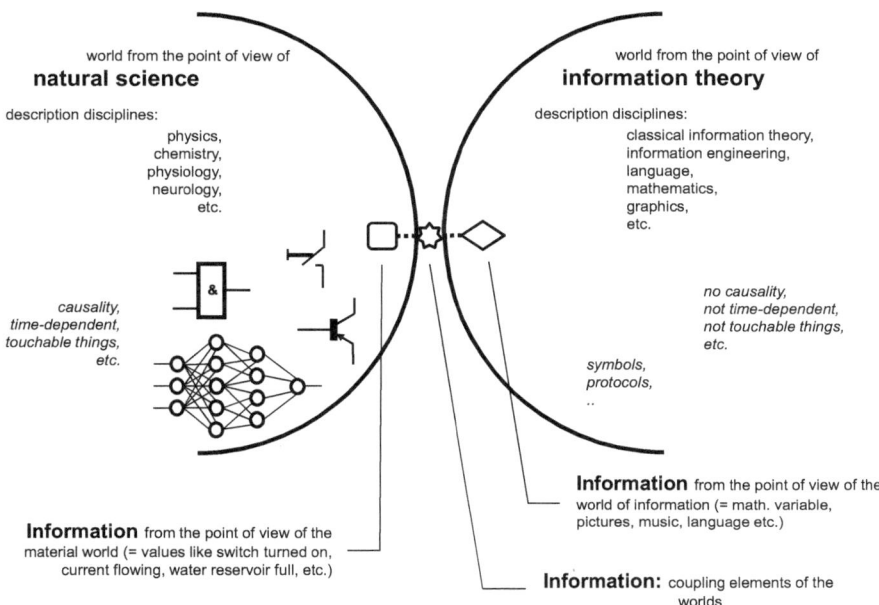

Fig. 5.2 Simplified representation of the separation between the two worlds and their respective methods of description

E. Shannon, such as the models and system languages for hardware and software developed in computer technology and computer sciences. In particular, I see the model that will be explained below, which connects the worlds of hardware and information theory. I see the continuous information *model of Georg H. Mealy*. All this is essential for understanding SiMA. Since information theory has grown into such a large field, this means that I will limit myself to information theory as it is relevant to SiMA.

I often get asked: What are, in contrast, the foundations of Norbert Wiener and Claude E. Shannon? Put briefly: Norbert Wiener formulated the functional theory of motion processes of objects on an axiomatic basis mainly in the field of *statistical* mechanics and many other issues. His work on automatic target control made him the "father of cybernetics". *Claude E. Shannon* defined on a *statistical basis* for the first time basic concepts of communication technology such *as information content, information loss* and *channel capacity* and introduced the *entropy theorem* for the definition of the information content of a source per character in this field. I do not want to go into the more precise formulations of these statistical relationships here. They are not necessary to explain the SiMA project. However, it is crucial to understand that these mathematical derivations are detached from physical description models (Steinbuch and Rupprecht 1967). Already for this reason the works of Norbert Wiener and Claude E. Shannon are important.

Applied to SiMA, from this explanation follows the following: If I start from the information α and β and connect them in some way, I get the information γ. Mathematically, this can be formulated as follows:

$$\gamma = f(\alpha, \beta), \tag{5.1}$$

which means that the information γ is a function of the information α and the information β. Which exact mathematical relation describes the function $f(\alpha, \beta)$ has to be investigated for each individual case, i.e. for each function. I will return[8] to this in Chap. 6.

Norbert Wiener and Claude E. Shannon already showed in such mathematical formulations that time plays no role in them. This means that it is, as Eq. (5.1) indirectly expresses, not subject to causality. *Causality* is part of physics. The behavior of time is a parameter to be considered only in purely physical terms. In the world of information, sequence information of processes and time are just sizes (information) like others, to be treated in the sense of Eq. (5.1). Nothing more. Whether these informational links are physically meaningful must be verified in the physical world. This relationship can be shown relatively clearly in the Mealy model.

Since the times of Norbert Wiener and Claude E. Shannon, there is still one point that poses a problem in the technical field. To this date no engineer has been able to define the term "information". As mentioned above, terms such as *information content*, *information loss*, and *channel capacity* have been introduced and thus recorded and described axiomatically. In contrast to these terms, "information" has to do with the *meaning of the content*, which in turn can be different for the sender and the receiver. To the best of my knowledge, only a few scholars in the humanities have attempted to work this out scientifically. Consequently, this will have to change with SiMA. Psychoanalytic models and brain research introduce valuation variables such as emotions and feelings. In this way, the concept of information can be brought into an axiomatic connection with *emotions* and *feelings*, which must ultimately lead to a natural scientific definition.

One can see that AI opens up new approaches.

5.2 Extended Mealy Model

Thanks to the work of leading figures such as Alan Turing, Konrad Zuse, Norbert Wiener and Claude E. Shannon, to name just a few, the world of information was defined in an ever more comprehensive manner with the help of mathematical tools.

---

[8] A detailed mathematical formulation of the functions f is of interest to the engineer when it comes to working out the function modules discussed in Chap. 6. In doing so, each function f will be different from the others.

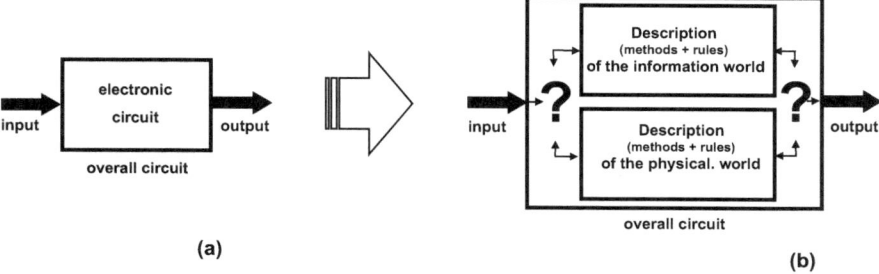

Fig. 5.3 Description of a circuit using two methods that display two different contents

While this world must be seen as independent of the world described by physical laws, but on the other hand, both worlds must be linked together. And it is essential to understand *how*.

I know I am repeating myself, but to understand the subsequent elaborations this statement is essential: We distinguish the different worlds only by using different *methods of description*, *methods of natural scientific approach*, etc. and by applying different *laws* to each of them. This is the crucial and fundamental fact. Another fact is this: If you want to simulate an *overall model* in natural science (which is what Fig. 5.3a) is supposed to represent), it is important to find the connection between the different worlds, otherwise both worlds exist alongside each other. As a consequence one cannot demonstrate through *experiments*[9] whether there are contradictions between them or not. A graphical explanation of this is shown in Fig. 5.3b, in which the overall circuit of Fig. 5.3a is split up. Philosophical answers are not sufficient for natural scientists. However, philosophers can help to question the logic and plausibility of the answers given by natural scientists.

For digital circuits, George H. Mealy[10] (1955) succeeded in solving the problem represented by the question mark in Fig. 5.3b. As a student I did not yet have the benefit of being able to study his scientific findings, although I only began my studies in 1970. I only studied individual electronic, digital circuits at the Technical University of Karlsruhe, which were then put together like Lego bricks to form an overall circuit (Steinbuch and Rupprecht 1967). But this bottom-up method has a serious disadvantage: Once the overall circuit has been completed, you do not know whether there are other circuits that fulfill the same function but contain fewer components and are therefore better (not only for cost reasons, but also because of reliability, energy consumption, etc). At that time, this led to competitive situations between us students, which is not a bad thing in itself as it sets a positive incentive.

---

[9]The term *experiment* refers to a wide variety of practical investigation to *validate* with clearly defined and established methods the theory, e.g. by simulations. They are a cornerstone of natural science.

[10]To derive the results described here from the work cited requires expert knowledge in the field. I therefore recommend that the interested reader first study Siegfried Wendt's book (Wendt 1974), which is already quite challenging.

But such a method is not good for the industry. Academic research had thus the goal to find a way to develop a digital electronic information system with the top-down method, e.g. on the basis of the *AND-*, *OR-* and *NOT-*functions,[11] which would *guarantee* a minimum number of building blocks. Georg H. Mealy found the mathematical solution. The bottom-up method, which I learned as a student, is what we computer engineers now derogatorily call "tinkering". In contrast to my student days around 1970, today every computer engineer knows Georg H. Mealy's[12] scientifically sound calculation method. I taught it for about 30 years by saying that Georg H. Mealy has developed a mathematical description of something that you cannot imagine, but with which you can do excellent calculations: He "tore apart" the circuit, which is matter, into the two worlds of *physics* and *information engineering* and coupled these two worlds via information paths (an interface for information). Today I can formulate it better. Georg H. Mealy drew from the fathers of information engineering such as Norbert Wiener and Claude E. Shannon. He used a black box to define and describe a space (a world, or whatever you want to call it) in which the laws of physics are applied, and a black box in which only information equations are applied. In the box of physics, the law of causality applies, time sequences are taken into account, physical quantities such as voltage, current and energy are taken as a basis and mechanical data are determined. The physical world is therefore quite easy for an engineer like me to understand in terms of presentation. In the world of information, it is generally somewhat more difficult. In this box there are no time delays, no causality. There, all variables are pure information variables, defined and described using symbols, for example, which can be formulated in terms of information engineering, i.e. according to Eq. (5.1). We have no sensory organs for this. It is a very abstract world that must be described and understood using mathematics. Physical laws must not be applied here.[13] The real genius of Georg H. Mealy was that he understood how *to connect* the two worlds (boxes) in a mathematical way, *namely via the information itself*, because it is this information that can be described in both worlds, in terms of *physics* on the one hand and in terms of *information engineering* on the other. This insight of his is ingenious and computationally non-trivial, which can be plausibly understood in Wendt (2013). Georg H. Mealy only had the digital world (Mealy 1955), i.e. electronic

---

[11]Following the logic of Sect. 3.1, these three functions can be considered as the basic functions (smallest units) of an information system, to which all other digital information systems can be reduced or mapped to.

[12]I hope it is not inappropriate to ask: Why did it take more than 15 years for electrical engineering faculties in German-speaking universities to include Georg H. Mealy's mathematical methods in their curriculum? Was it because it was a paradigm shift?

[13]From a psychoanalytic perspective, one can associate the primary process at this point, since in it the information is unconscious and does not conform to physical laws (List 2009, p. 74; Freud 2020b, p. 287). Physical laws, as well as causality, play no role there. A psychoanalyst should therefore be able to spontaneously embrace this model. When I, as a non-psychoanalyst, realized that *Sigmund Freud describes the primary process as a pure information system*, just as Georg H. Mealy does with the box of the information world, I was highly enthusiastic because I then knew that the principle, underlying the SiMA project, must be the right way to go.

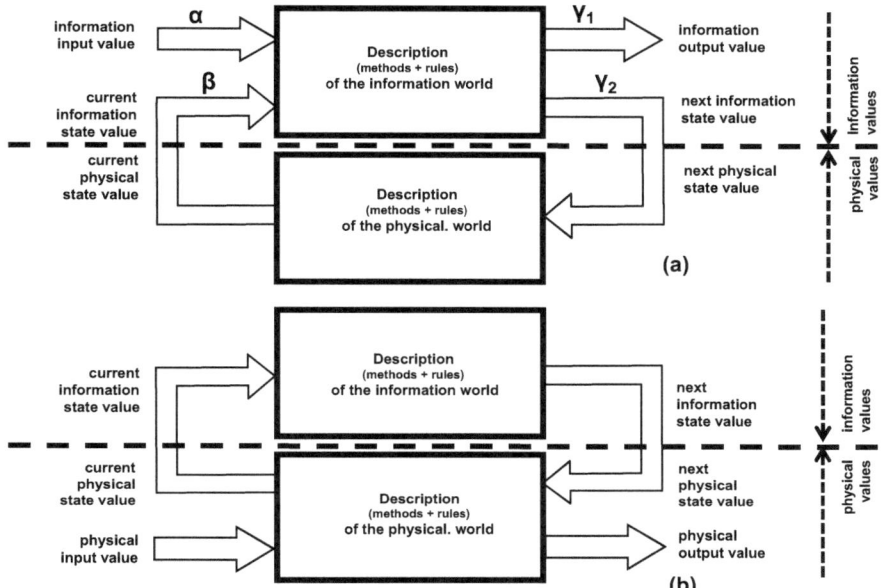

Fig. 5.4 Mealy-Model (one can discuss for a long time whether one should speak of worlds or spaces. In principle, both terms are associated with something material, which is fundamentally wrong, because we are working with mathematical "spaces" or "worlds" here. These have little to do with our three-dimensional conceptions. However, I do not know any better terms, so please forgive me if I use the terms "spaces" and "worlds" synonymously and, above all, perhaps too informally). (**a**) *Input and output variables are information variables,* (**b**) *input and output variables are physical variables*

circuits, in mind. I will come to a generalization later. The principle can be illustrated and applied in different ways. I have abstracted this graphically and reproduced it in two variants in Fig. 5.4a, b.

Why are there different illustrations? This is mainly due to mathematical and application specific reasons. The model depicts an abstract space in which information variables are processed and computed above the dashed line and physical variables are processed and computed below the dashed line. The area outside the two black boxes is undefined, only the arrows outside the boxes indicate the inputs and outputs at which the information and physical variables, respectively, must be considered. Figure 5.4b is the less abstract representation. The input variable of the model is a physical variable, i.e. it is physically adapted to reality. However, the model is more difficult to derive mathematically. In contrast, Fig. 5.4a is more abstract, the input variable is an information variable. The model is therefore easier to describe mathematically and therefore easier to explain. However, it requires a trick to bring it together with our physical world. Why? For a simple reason: How can an information variable with no physical basis enter the model? It cannot! Every piece of information needs a physical carrier, for example acoustic or electric waves.

But it is not important to deal in detail with this problem in order to understand SiMA. The only thing that is important for a basic explanation of Georg H. Mealy is to know that both models can be converted mathematically into each other. This allows me to develop further explanations based on Fig. 5.4a. Here, let us ignore Fig. 5.4b and follow only the various variables α, β, γ_1 and γ_2 plotted on their paths in Fig. 5.4a.

Output variables are calculated according to Sect. 5.1 using the equation

$$\gamma = f\,(\alpha,\ \beta). \tag{5.1}$$

Applied to the upper box of Fig. 5.4a, this means that the input variables are the pieces of information α and β and the output variables are the pieces of information γ_1 and γ_2. The letter f stands for function. The equations of Fig. 5.4a can therefore be formulated as follows:

$$\gamma_1 = f_1(\alpha, \beta) \quad \text{and} \tag{5.2}$$

$$\gamma_2 = f_2(\alpha, \beta). \tag{5.3}$$

This formalism means: α is the information input variable, γ_1 is the information output variable of the complete model. The information input variable α comes from another information system that is not shown in Fig. 5.4a, and the information output variable γ_1 is transferred to an information system that is also not shown in the figure. γ_2 represents the information state variable that is required in the lower box (*laws and description methods of the physical world*) for the calculation of the future (physical) state variable. What does this mean? In the lower block, the *information* arriving from above is considered as a *physical variable*. The necessary physical laws, parameters, time constants, etc. are described in this box. The output variable β from this block is therefore also a *physical variable*, which is then used in the upper block as an *information variable* in the equations of the information world. At the transition—represented by the horizontal dashed line—from the world described in terms of information engineering to the world described in terms of physics, γ_2 is therefore reinterpreted, and the same applies to β at the transition from the world described in terms of physics to the world described in terms of information engineering. And reinterpretation means that β and γ_2, for example, have the unit milliseconds, volts, or amperes in the physically described world, whereas in the information engineering world they are purely symbolic quantities without units.

But what do these two equations mean and what do f_1 and f_2 stand for? As already mentioned, they state that the output variables γ_1 and γ_2 are each a function of the two input variables α and β. These functions f_1 and f_2 are the result of an electronic circuit, which, for example, consists of *AND*, *OR*, and *NOT* functions, as shown in Figs. 3.11–3.13. In other words, the entire upper information black box of Fig. 5.4a contains two functions (circuits) that are represented graphically. In other words, the entire upper information black box of Fig. 5.4a contains two functions (circuits), graphically represented as circles in Fig. 5.5a. They receive the input variables and

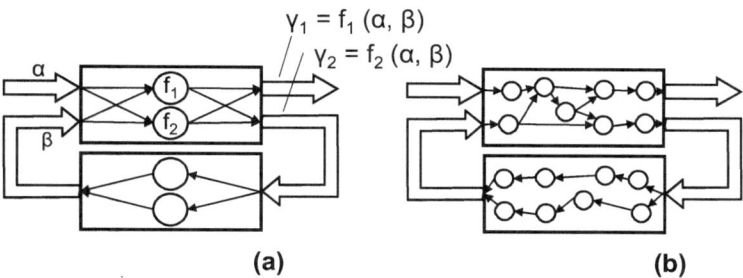

Fig. 5.5 The principle of Mealy following the illustration in Fig. 5.4 (**a**) with the representation of the two functions f_1 and f_2 in the informationally described box, (**b**) f_1 and f_2 split into many functions in the informationally described box; the structure in the physically described box is not of interest here

generate the output variables γ_1 and γ_2. In the upper black box, these are symbolic information variables. The physically associated variables such as currents, voltages, times or frequencies etc. are specified in the lower black box.

The two functions f_1 and f_2 of the information box in Fig. 5.5a are usually composed of sub-functions, which leads to the likewise highly schematized exemplary Fig. 5.5b. The sub-functions can then be further subdivided down to the elementary digital, logical circuit elements such as the basic functions AND, OR, and NOT. Again, their physical equivalents in the lower box of Fig. 5.5b must be taken into account. Each logical circuit function is therefore represented twice, once in the information engineering box and once in the physically described box. Note again:

> The information box and the physical box only represent two description methods for one and the same (material) unit.

In the field of chip design or chip computation (general circuits and microprocessors in particular), the application of Mealy theory, as shown in Fig. 5.4a, has prevailed over the representation shown in Fig. 5.4b. The corresponding Mealy equation systems are directly or indirectly integrated in all design languages such as VHDL, Verilog or SystemC. They guarantee an optimal circuit for the hardware designer according to the top-down principle.

However, the structure of Fig. 5.4a has a major disadvantage. In principle, it assumes that the input is described purely as information. A certain physical carrier of the information is assumed. However, it is omitted from the description for the sake of simplicity. If such a system from Fig. 5.4a is coupled with another subsequent system, the result is Fig. 5.6a. Two coupled Mealy models are shown graphically. The question mark is intended to indicate that the transfer from one model to the next can only take place via matter (information carrier), which consequently has to be agreed upon in addition. It is not included in the picture. In the field of chip design (development of electronic circuits), standards have been agreed upon, which is not difficult if one considers that the developer's task is to adapt the basically asynchronous world from the outside to the synchronous world

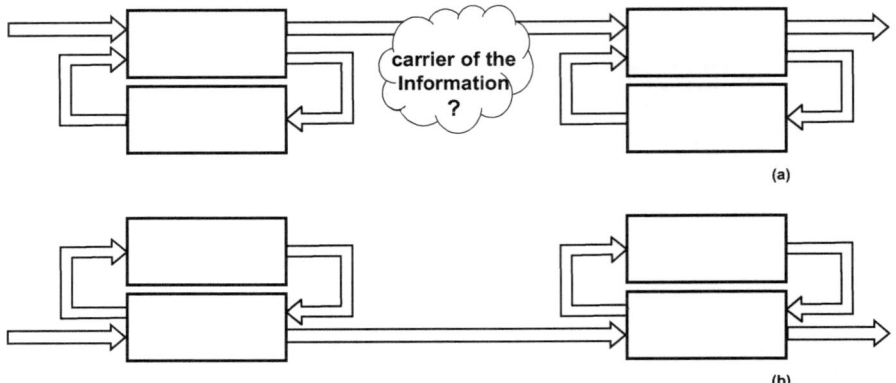

Fig. 5.6 Coupling of Mealy models. (**a**) *Application in the field of chip design,* (**b**) *application in the field of technical communication systems*

from the inside. So, the step to consider information as a physical quantity is not a big one. On the other hand, the mathematical handling of the overall model (the Mealy equation system) is simpler compared to the structure of Fig. 5.4b.

In practice, the development of communication protocols in the field of computer engineering is completely different (Zimmermann 1980) to that of chip development. There, a wide variety of information carriers are used: metallic cables, radio, infrared or ultrasound, completely different electrical values such as voltages and currents, and many others. The technology used to connect the circuits plays a central role. It cannot be standardized for cost reasons—there are far too many possibilities—and therefore cannot be ignored in plans. This leads to the illustration in Fig. 5.4b. Of course, the underlying principle is the same as in Fig. 5.4a, except that the inputs of this structure flow into the lower block as physical variables. In the same sense, the output variables of the entire structure are also physical variables. The concrete mathematical description of Fig. 5b is therefore different from that of Fig. 5a. When several models are coupled, the representation leads to Fig. 5.6b. Unfortunately, the coupling between the two worlds (the physical world and the information world) becomes more complicated, which I will discuss in the next chapter with regard to the extended Mealy theory.

But before I get into that, I want to point out two things. First, those interested in technology should know that Georg H. Mealy distinguished between asynchronous and synchronous circuits, as I mentioned earlier. Asynchronous circuits are circuits whose components are not based on a common clock. *The Ψ-organ is a prototypical asynchronous system.* Each neuron fires when its threshold is reached, independently of the thresholds of the other neurons. In contrast, synchronous circuits are based on a central clock. Today's computers operate synchronously. The clocking of the circuits in a computer always occurs at the same time intervals and triggers another information step, i.e. information is stored, transferred, and/or processed in all circuits at the same time. For the model in Fig. 5.4a, for example, this means that information, present at the input on the right in the lower block—i.e. in the physical world—, reaches the output on the left of the physical block only when the common clock of the box and thus of the entire circuit allows it. No register, no flip-flop in it switches independently of this common clock. In an asynchronous circuit, on the other hand, each physical signal is passed through the lower block immediately, but experiences a time delay due to the laws of physics, which depends on the components used.

The calculation of asynchronous circuits[14] is time consuming and relatively difficult. I taught it for many years in the hope that science would develop simple tools (algorithms) for it. But so far there has been no breakthrough. However, asynchronous circuits offer many advantages. Hopes for such a breakthrough have risen repeatedly, but in the end the cost of developing and testing such circuits has always remained high, which is why industry is not interested in asynchronous circuits. However, there are exceptions, namely where there is no other physical possibility. For example, an asynchronous circuit must be used at the input of circuits whose input signals must first be synchronized to the circuit's internal clock. In such cases, you have to bite the bullet. But then you try to get by with as few, easily manageable components as possible. This approach is a must in elec-tronic circuit design today.

The bottom line is that whether synchronous or asynchronous circuits are used in SiMA plays a minor role. However, the issue should not be ignored. It mainly concerns hardware, less information engineering. Where it has an impact on infor-mation engineering, the effect must be specifically investigated, but this should only be a question for the computer engineer. This means that the topic of synchronous or asynchronous circuits must not have any influence in a simulation, i.e. it must not affect the experiment. The computer engineer must always keep an eye on it. I will come back to this briefly in Chap. 6.

The second thing to be noted is crucial for the following chapter: In his first publication, Georg H. Mealy (1955) developed—as I have already said—a system of equations on the basis of which in principle all (i.e. both asynchronous and synchro-nous) electronic digital circuits can be calculated. As I explained above and will explain in more detail in the next chapter, his model—which I present in Fig. 5.4b—can also be used for communication systems, but its mathematical formalisms cannot be adopted directly. Mealy's principle has to be adapted, i.e. generalized. Hence the term *Extended Mealy Model*. This can be done to the same extent in a subsequent step for modeling a neuronal network, which will be discussed in Chap. 6. First, I simplify Fig. 5.4 and arrive at Fig. 5.7a for an electronic circuit as an example and (following the fundamental assumption that the neuronal system is the hardware of the Ψ-organ) I finally arrive at Fig. 5.7b for a Ψ-organ. The equivalence of the two

---

[14]This topic may not be very interesting for psychoanalysts, but if you want to simulate the Ψ-organ, this topic has to be addressed. This is because our Ψ-organ is an asynchronous system from a hardware (neurological) point of view, but in SiMA we are forced to base the hardware on a synchronous system. This must be taken into account from an information-theoretical point of view.

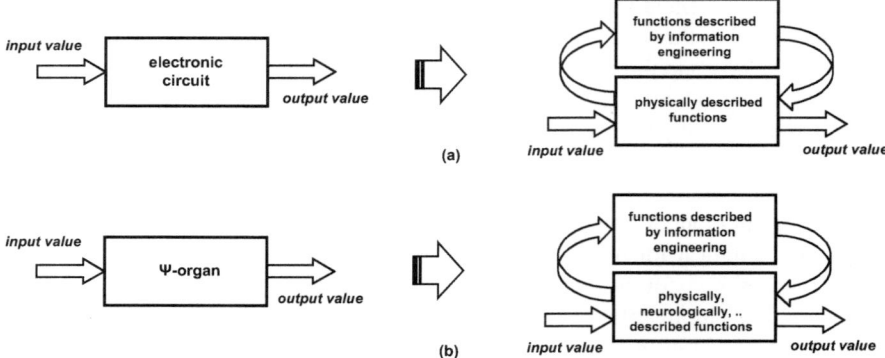

Fig. 5.7 Basic idea for modeling the neuronal system or Ψ-organ according to the principle of Georg H. Mealy. (**a**) *Generalized Mealy principle*, (**b**) *the Mealy principle applied to the Ψ-organ*

figures to the right of Fig. 5.7a, b is intended to highlight the point made in Sect. 3.1: From the point of view of information engineering, the models essentially do not differ with regard to Mealy's theory, only the concrete mathematical descriptions (formulations) must be approached differently, but I do not want to go into this here.

Figure 5.7b will be used in Chap. 6 to develop the SiMA model, i.e. the model of the Ψ-organ in the form of a joint neurological, psychological, and psychoanalytic description. First, however, let us turn to the model of technical communication, which can be derived from Fig. 5.4b and which can be imagined as a network via Fig. 5.6b. It can be seen as an *extended application of the Mealy principle*, and it shows how difficult it is to get to grips with such models. This also explains why it is so difficult to convince people who think differently to familiarize themselves with these mathematical tools, even though there are no other alternatives. People just like to stick to their own way of thinking. As a classically mathematically oriented electrical engineer, I realized this very late.

5.3   Layer Model

I mentioned it before: Until the mid-1980s, most of us communications and information engineers lived in a perfect world of mathematics that people like Norbert Wiener and Claude E. Shannon had formulated so clearly for us. And as late as the mid-1990s, mathematically oriented hardliners in communications engineering were still telling me that anything to do with communications *protocols* had nothing to do with natural science.[15] The basis of communications engineering is electrical

---

[15] This has happened more than once. I have often been told that the sciences of communications engineering can only include subjects that include at least the mathematics of Claude E. Shannon in their formalism. And on what mathematical formalisms of communications engineering are

engineering, and therefore physics, and only mathematics is the accepted tool for electrical fields. Communications protocols have nothing to do with that. They were not familiar with or interested in the work of Georg H. Mealy. In contrast, relatively early on, in 1977, a committee was formed in ISO that understood that communication between computers requires appropriate rules, called *communication protocols*. When humans communicate in everyday life they do this intuitively.[16] In what order and at what moments can communication take place? How should it be done? What needs to be taken into account? And above all, how can these protocols be described and formulated scientifically? Doubts arose again and again: Is this really still science? There was massive resistance from the electrical engineering community. For this reason, the special ISO committee that took up the challenge was composed mainly of scientists from the fields of computer engineering, computer science, and mathematics. They were very familiar with the theories of Norbert Wiener and Claude E. Shannon. But they also understood that a new era had begun. One just had to realize it. The new way of thinking posed problems for many engineers. They had to understand that a distinction had to be made between physical and information processes. This means that the laws of physics can only be applied to the necessary physical information carriers, while the communication process itself, i.e. its protocol sequences, cannot be described by the laws of physics and the mathematical formalisms of classical electrical engineering. *New approaches were needed.* And the experts had various communication processes in mind. But how did they come up with these ideas and how did they reach an agreement? They had to solve the following problem: In mathematically described, physically formulated laws, compromises were unknown. The result was either right or wrong. Yet, in stark contrast to this, communication protocols do not allow for such clear categorization. In most cases, different solutions are conceivable, all of which have their advantages and disadvantages.

At my company, I met a computer scientist who was a member of this committee. His stories were always exciting for us mathematically oriented engineers. Later, I also met other scientists who, here and there, especially at conference introductions, would tell stories, mostly with amusement, about the years-long process of developing this first international standardization, which was ultimately hard-won. A specific communication process is always only one possibility among many others. But the goal was to develop *one* binding standard, otherwise the industry would have no chance of developing compatible devices and, above all, a unified network. This would have resulted in isolated solutions, a nightmare for computer engineers. A

---

protocols based? They are not! Therefore, it is not a natural science. This is how it was stated to me. When I first heard this, I dismissed it as a colleague's abstruse idea. But then such statements were repeated. And it is interesting that even today such views are held by renowned scientists. This gives you a good idea of how narrow-minded some scientists are, even though they are capable of enormous achievements in other fields.

[16] In parallel to this development, the TCP (Transmission Control Protocol) was developed, which I will not explain in detail here, as it was not intended to be generally defined for different communication processes, unlike the one discussed above.

computer only makes sense if it can "talk" to others, i.e. if the communication procedure is based on *one single* standard.[17]

On the other hand, the following problem arose: If the standard was so prescriptive that every company had to build the same, structurally identical components, price competition between companies could only be handled through production costs and thus ultimately through wages, which would destroy smaller companies. Thus, a compromise had to be found for several demands, otherwise no solution would have been found in ISO. And this compromise was finally reached: The communication model should not be a design specification, but only a tool, an aid, a *working template* for the developer who wants to design a communication system. It should show them the *possible functions* of communication. If they used this *working template* as a basis, they could refer to this ISO/OSI model.

So how did the experts on the committee come up with the name for this communication model, this *working template*? On the one hand, they were annoyed with CCITT because by this institution they had not been spoken to as equals and had even been rejected outright. On the other hand, they were grateful to ISO. There they had been able to bring their ideas to the ISO committees, and decisions had been made accordingly. This should be reflected emotionally in the name. They reversed the order of the ISO letters and came up with the name *Open Systems Interconnection* (OSI). Today it is commonly referred to as the ISO/OSI reference model. However, some people prefer the name seven-layer reference model.

And how did the experts on the committee arrive at their ideas? It was a relatively complex process, as they first had to agree on all the requirements that the future devices (and these were ultimately computer-based devices) would have to meet. Even more difficult, the specification to be developed had to be generally applicable to the various communication principles used for computers. Many called the result of these requirements the "eierlegende Wollmilchsau" (an expression that literally translates as "egg-laying wool-milk sow" and is used in German to describe a desirable but unrealistic or hard to find all-in-one solution).[18]

---

[17] Again, these discussions may sound completely uninteresting to psychoanalysts. However, it must be remembered that this process represented a tremendously revolutionary and radical change in the way scientists and engineers worked. Engineers had to bite the bullet and not just think outside the box. They were forced to familiarize themselves with the new field and learn completely new methods. All experts have to go through a similar process if they want to work out the connection between the nervous system and the psychic apparatus for the Ψ-organ in a natural scientific way. That is why I am going into so much detail. It was the same giant step for the engineers (e.g. for me) that neurologists and psychoanalysts must now take if they are to arrive at a unified model of the nervous system and the psychic apparatus.

[18] The advantage of such an "egg-laying wool-milk sow" is obvious: (1) Everyone is made aware of how many functions a communication could contain. (2) You can pick out the functions that you consider relevant for the respective case. (3) There are many degrees of freedom in development. (4) Every developer can use the model as a reference list so as not to forget any functions. (5) Comparison with non-technical communication principles is made easy as the explanations of the ISO/OSI model are based on a broad basis.

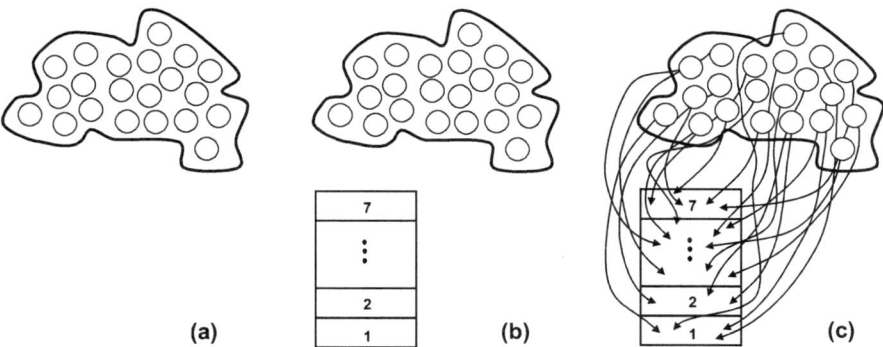

Fig. 5.8 Assignment of defined communication functions in the ISO/OSI model with hierarchically ordered layers. (**a**) *Collection of all communication functions,* (**b**) *communication functions and hierarchical layer model,* (**c**) *assignment of the communication functions to the seven layers*

The first step was to collect all functions related to communications, hardware functions as well as functions related to the laws and regulations of information. All these functions are shown as circles in Fig. 5.8a. It was agreed without much difficulty that the ISO/OSI model should be hierarchically layered rather than democratic, since it had to work quickly and in a simple manner (Fig. 5.8b). However, the problem that arose was: How many *layers* should be defined, 3, 10, or even more? I have never found a documented account of how this came about, but I have received several reports from people who were involved. So, it may or may not be true, but the explanation I heard made sense to me. I have experienced similar compromises in other committees such as ISO. Over 100 different nations from different cultures were involved in this ISO/OSI committee, so there were a lot of different proposals. However, in the evening, over dinner and good drinks, the only possible compromise was seen in the special—some spoke of the "sacred"—number 7, which is why the ISO/OSI model, as shown in Fig. 5.8b, has seven *layers* (not to be confused with the *levels* of the abstraction model in the next chapter).

The last step in forming the seven-layer model was then merely a task of diligence: All the hardware and information functions found had to be assigned to the seven different layers, as shown in Fig. 5.8c. How this was done in detail is no longer of interest for the SiMA project. I want to mention just two crucial aspects.

Aspect 1 concerns layer 1: As in Mealy's theory, the lowest block should include everything that comprises the pure physical laws as well as the necessary specifications.

Aspect 2 concerns layers 2–7: All layers above layer 1 are dedicated to pure information theory and information engineering considerations (regulations/protocols).

These two aspects have important implications. The interface between layer 1 and layer 2 as well as the interface of layer 7 "upwards", i.e. to the (abstract) object above, are of outstanding importance (Fig. 5.9). In the interface transitions of the interfaces between layers 1 and 2, the information contained therein is treated as

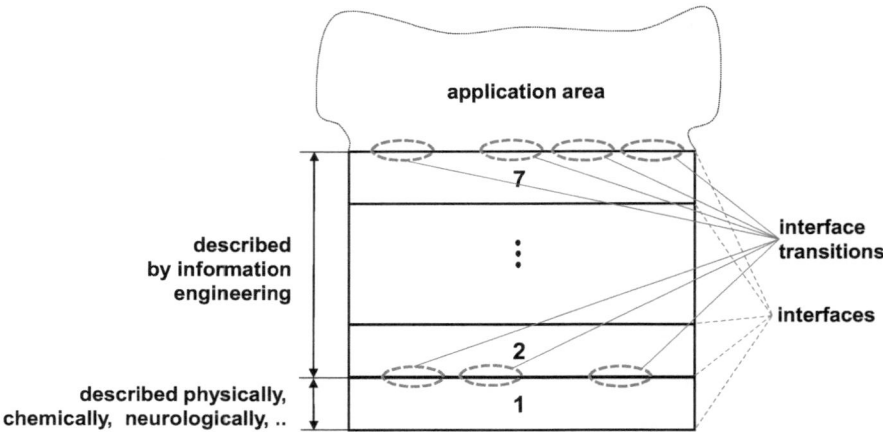

Fig. 5.9 ISO/OSI model and its interfaces. *The application is no longer included in the ISO/OSI model*

physical sizes as far as it relates to layer 1, and as far as it relates to layer 2, it is considered and calculated from an information engineering point of view, i.e. using the methods and laws of information engineering. Layer 7 has the sole task of adapting the information given to the superordinate system to the requirements of that system and, conversely, of adapting the information coming from that system to the ISO/OSI model. The superordinate system can be a computer or any other kind of device. The interface transitions in the interface between layer 7 and the overlying unit therefore depend on the requirements of that unit.

As far as the SiMA project is concerned, the following crucial conclusions can be drawn directly from the development of the ISO/OSI model.

1. *The principle of Georg H. Mealy has not been distorted in the ISO/OSI model, it has only been extended.*
2. *Mealy's two-layer model can be subdivided into as many layers as desired, depending on what is the most efficient solution.*[19] *It is only necessary to make sure that the layer types described by physical, mechanical, chemical, etc. laws are not confused with the layers defined solely by information engineering laws or protocols. The lower layers are therefore always the physically oriented layers.*
3. *The interface transitions between the layers must be clearly defined.*

---

[19]The ideal number of layers depends on several criteria. For example, in the engineering domain, the following can be said: The more layers you define, the longer it takes for information to travel through them; the fewer layers you define, the less comprehensible the device becomes, which increases the cost of development and, above all, maintenance. With respect to SiMA, other specifications apply, which are discussed in Chap. 6.

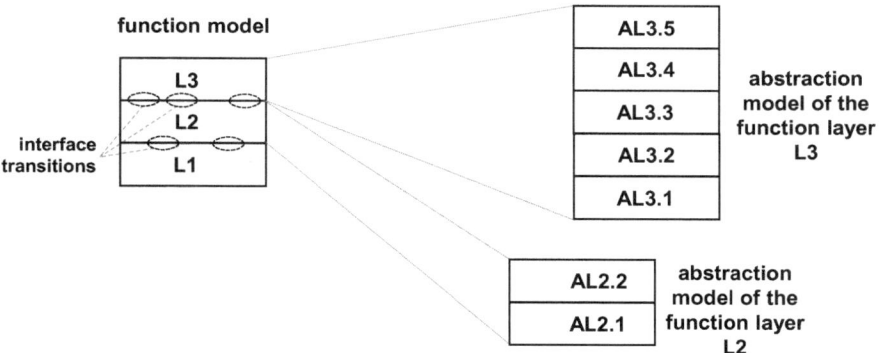

Fig. 5.10 Abstraction models for the functional layer L3 and L2 *(Lx: functional **layer** x, ALx.y: **level** of abstraction x.y)*

If you want to develop extensive, hierarchical models according to ISO/OSI, you run into a difficulty: The individual layers can become very extensive, which makes them confusing and leads to development errors. To avoid this, the abstraction model has been introduced, which will be the subject of the following chapter as I consider it essential for SiMA.

5.4   Abstraction Model

The individual layers in a functional model, such as the ISO/OSI model, contain different functions (Fig. 5.8c) with different tasks. For example, layer 1 describes physical functions, while the top layer, i.e. layer 7, describes functions that have the task of adapting the model to devices which are arranged hierarchically above it. If the layers contain many functions that require extensive description, there are two options. One option is to create subordinate layers. This solution creates new layers between these sublayers and thus creates new technical requirements. However, technical requirements are not the subject of SiMA and should not be of interest to us here. The other option is to consider whether a *hierarchical abstraction model* should be developed for *each* layer of the layer model. This does not change the layer in question, i.e. it does not modify the functions, it leaves them as they were originally defined and only describes them in different levels of abstraction. How can you imagine this?

Consider a simple technical example. An information system is to be built with three functional layers. Of course, the lowest layer is the physically described function. In practice, this is usually not described by an abstraction model. There are more efficient methods for this, which will be discussed briefly in the next chapter, as it only touches on SiMA in passing. Let us focus on layers 2 and 3 and call these functional layers L2 and L3, as shown in the left part of Fig. 5.10. Of

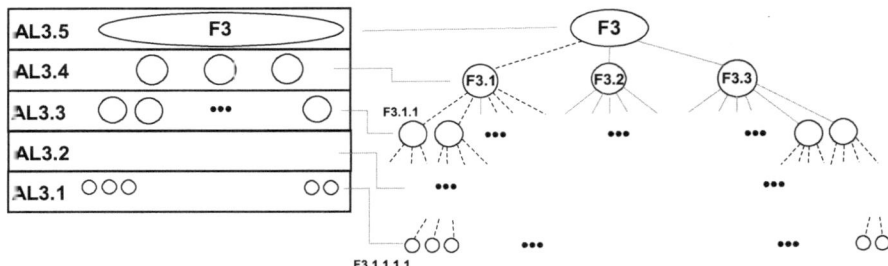

Fig. 5.11 Breakdown of the abstraction model of the functional layer L3 *(ALx.y: level of abstraction x; Fx.y: function x.y)*

course, the interfaces between the functions, i.e. the information transitions, must always be taken into account, but this is not important at the moment. The question should be: How much effort is required to describe the two layers? Let us assume that layer L2 has a relatively simple structure, while the functional layer L3 requires a high level of description. Thus, different numbers of abstraction levels can be assigned to the abstraction models.[20] As shown in Fig. 5.10, I have assigned two abstraction levels to layer L2 and five abstraction levels to layer L3. What does this mean?

Consider the somewhat more complicated case of layer L3, which is shown in Fig. 5.11 only as an abstraction model. For simplicity, the information transitions are not included. In the topmost abstraction level AL3.5, the function of the functional layer L3, labeled F3, is described as a whole (symbolized in Fig. 5.11 on the right as a circle in the topmost abstraction level). The function F3 is thus described in a highly abstracted form at this level of abstraction. All the details of layer L3 are omitted. The function description is summarized in F3. In the abstraction level below AL3.4, this function F3 is divided into three sub-functions F3.1, F3.2 and F3.3, i.e. it is described in a less abstract way. At the next lower level of abstraction, the sub-functions are divided again, and so on, until the smallest functions are obtained at the lowest level of abstraction AE3.1 (in Fig. 5.11 with the many functions[21] F3.1.1.1.1, F3.1.1.1.2, F3.1.1.1.3, ..., F3.2.1.1.1, F3.2.1.1.2, ... up to F3.3.x.y.z), which can be "easily" implemented, e.g. in software units.

This may sound easy to implement. But in practice, it can be challenging. After all, the software engineer sees many, many "small" functions that all somehow work together, i.e. are interconnected and interact with each other in some way. This can get quite complicated. First, the term "small" in the phrase "small functions" must be

---

[20] According to the Mealy principle, the use of the terms *layer* and *level* is arbitrary. From the point of view of the ISO/OSI model, which generally uses the term *layer* as a standard, a different term should be used for the abstraction model in order to make a clear distinction. Therefore, I decided to use the term *level* in SiMA.

[21] In SiMA, three functional layers were selected and nearly 50 functions were defined for layer L3 (the psychic apparatus) (see Chap. 6 for the explanation).

seen in relative terms. Only from the perspective of the hierarchical abstraction model can the functions at the lowest level of abstraction be described as "small". We have never been able to complete the development of such "small" functions with the scientific effort of a single dissertation.[22] On the contrary, the deeper we delved into the subject, the more it became clear that we had to go much deeper. In general, I see functions in the SiMA model as "never-ending stories". It is like physics: The more physical units you break down, the more questions arise—an inherent fact of natural science. Moreover, with the SiMA project, we are at the very beginning of a new scientific field in which we are trying to model and record the functions of the Ψ-organ. However, we are very confident that our approach is correct, as the experiments so far have not shown any contradictions.

Second, there is the challenge posed by an extensive network of functions, such as that obtained at the lowest level of abstraction (in Fig. 5.11, this is AL3.1): To describe all the functions of the L3 layer and their behavior. The functional layer is therefore described at this level of abstraction by many "small" functions. To evaluate this network, a correspondingly large number of experiments must be performed. And in the case of SiMA, these experiments can only be designed and evaluated by the relevant experts in psychoanalysis. The engineer can only play a supporting role.

5.5   Challenges of Information Theory

The functional modeling of the Ψ-organ, the step that can make the bridge between the neuronal and the mental part comprehensible—the ultimate goal of SiMA— demands new ways of thinking, ways of understanding information theory. The part of information theory that I have presented so far is only a small part of today's very extensive knowledge. I think of other helpful tools in this area that have been developed over the last few decades. The Y-diagram of Daniel D. Gajski and Robert Kuhn, which substantiates the extended principle of Georg H. Mealy in a very comprehensible way, should be mentioned in particular. As head of my institute at the Vienna University of Technology, I had the pleasure of working with Daniel several times when he was a visiting professor at our institute. Although he is also a strictly hardware-oriented person, he knows how to keep the two worlds of physics and information engineering strictly separate. He, in fact, does more than that, and this is his real achievement: His Y-diagram differentiates in detail between different approaches. He not only separates the physical and the information theory like Georg H. Mealy, but also introduces more than *two ways of describing* them, and thus divides them into more than two layers.

---

[22] It is important to note that the average time to complete a dissertation in SiMA was about 5 years. This was due to the need for a solid grounding in the various topics of computer engineering, computer science, and psychoanalysis.

I also think of the design language SDL (Specific and Description Language). It is ideal for developing and programming the abstraction model without contradictions. I would also like to explain the terms *data model* and *behavioral model* in more detail, and especially their differences. However, I would inevitably end up getting too technical, which might put some readers off. However, there is no getting around the part of information theory that I explained in the previous chapters; you have to understand it at least to some extent if you want to immerse yourself in this world.

Perhaps I succeed in resolving the resistance that some readers may have towards the stringent differentiation between physical descriptions on the one hand and informational descriptions on the other. I have almost always been confronted with this when I have lectured to psychoanalysts, psychologists, and physicians. Sigmund Freud's wish was to embed psychoanalysis in the natural scientific way of thinking (Schülein 1999, p. 35), but he did not succeed. From today's point of view it is obvious why he could not possibly succeed. On the one hand the technical theory of information could—unlike psychoanalysis—only be developed gradually and slowly with electronics. People understood how to assemble small electronic components in order to study their behavior in detail. We engineers really learned to deal with information theory only slowly—just like small children learn. We learned to understand and describe its natural scientific laws brick by brick, using small electronic circuits. We painstakingly learned to distinguish between information theory and physics, a colossal process of abstraction, probably the most important step. We slowly developed a "gut feeling" for this knowledge. Sigmund Freud, on the other hand, immediately had to deal with what is now considered a biological "supercomputer"—the human Ψ-organ. He had to develop methods that would help him understand and describe this concentrated information system as such (also in order to be able to work clinically). We computer engineers must recognize that, despite these enormous difficulties, he anticipated our conceptual definitions that are necessary for the development of hardware and software; e.g. with the development of the structural model, he anticipated the distinction between a *functional* and a *behavioral model*, which requires a complex theory of information. When I realized—"grasped"—this fundamental insight, I believed that it should be easy to communicate it to psychoanalysts, psychologists, and physicians. I believed that I would be able to reach them in their field of knowledge, i.e. where they stand scientifically. I was very much mistaken! I was also mistaken in thinking that this new insight would be easy to explain to engineers. I tried it for the first time at the WFCS 2000 (Dietrich and Sauter 2000) and was met with great incomprehension.

I see three main hurdles to overcome. First, the two scientific fields use completely different conceptual frameworks and ways of thinking, rely on different findings and methods, cultivate different rhetorical styles, and, above all, have completely different histories and traditions. To break away from all this and embrace the abstraction of Georg H. Mealy is asking too much, even for die-hard natural scientists who have nothing to do with information theory. It can only be done step by step. How long did it take for electronics to be accepted by doctors in hospitals? It took decades. How long did it take for computer technology to be seen as more than just something dangerous? And this is all just about acceptance. SiMA, on the other hand, is about much more. Both fields of knowledge have to be understood to some extent. The conceptual models of one world cannot be reconciled with the conceptual models of the other world. They are not compatible without the understanding of Georg H. Mealy. As a scientist, you have to understand psychoanalysis, and as a psychoanalyst, psychologist, and physician, you have to be able to accept the new findings of information theory—and that means accepting the world of mathematics with its abstractions without any reservations.

And this brings us to the second hurdle, which I think is crucial. Black holes in galaxies can, for example, only be understood mathematically through knowledge of the singularity, otherwise they remain a mystical concept.[23] The same goes for the theory of relativity. And in the same way, the bridge between neurology and the psychic apparatus can only be understood through a construct like that of Georg H. Mealy, if we want to describe it in scientific terms. We need such a mathematical model. As a scientist in computer engineering and information theory, I know of no other.

But everything has its limits: In their obvious eagerness for a solution, many well-known scientists resort to correlation methods in order to mathematically explain the connection between the two worlds. *This is the wrong approach.* Correlations can only be applied *within one* world, not *between* different worlds. We have defined physics within physics and information theory within information theory. They are to be seen as independent spaces of (axiomatic) definition. And I think—even if it sounds presumptuous—that the problem of these scientists is that they have not yet sufficiently dealt with mathematics and where its limits lie. Even Baruch de Spinoza, whom I greatly admire, completely overestimated mathematics (Mauthner 2015). Mathematics is only a *tool*. Mathematics helps to give an exact description. It does not replace physics, chemistry, psychology, or psychoanalysis, and it does not prove their correctness. In contrast to mathematics, the methods and laws of the natural sciences must be evaluated by experiments.

And there is a third hurdle that I keep noticing in conversations and letters that should not be underestimated. For example, a mathematician once wrote to me: "For me, the brain is not a computer," as Marianne Leuzinger-Bohleber put it in (Leuzinger-Bohleber 2008, p. 39). I argue that the thinking of Cartesian dualism developed by Descartes, that matter and mind are two different substances, persists in us despite the Enlightenment work of Baruch de Spinoza and others, because it underpins the religious ideas by which we have inevitably been shaped. This third hurdle must be seen as very powerful because it builds up enormous defense forces within us.

---

[23] And I claim that most people who talk about black holes have only mystical ideas about it. This is because they have never studied the mathematics of the singularity, which is necessary to understand it.

Chapter 6
Development of the Ψ-Organ Model (SiMA Model)

In SiMA, the Ψ-organ is regarded as an information system to be described in natural scientific terms. This requires a model that must be evaluated through experiments. What are the aims of these experiments? I want to be able to answer questions such as: What characterizes human beings? Into which functions can the Ψ-organ and, above all, the human psychic apparatus be broken down? How is human behavior brought about by these functions? How do human individuals differ from one another? Why will we probably never be able to simulate a living human being, and why does the simulation of the Ψ-organ still make sense? Finally, I must be able to demonstrate how I can experimentally evaluate the functional model of the Ψ-organ developed in this way, in order to then integrate the designed model into a robot in a further step. This in turn means that real-time requirements[1] must be taken into account—and the robot is then no longer a model in the natural scientific sense, but a concrete subject with emotions and feelings that exhibits concrete (individual) behavior. But what kind of emotions and feelings are these? Human emotions and feelings? Answers to all these questions will follow in this chapter.

Now, that I have provided a comprehensive overview of the theory of technical information and its tools, all the prerequisites are in place for me to finally delve into the topic of the development of the SiMA model. However, I must not forget to first define the crucial limitations of model development, as is customary in the natural sciences. Thus, I can make the reader aware of where the boundaries lie.

The goal is to model and simulate the human Ψ-organ. From the point of view of neurology, we mean all the neurons (Brainin et al. 2004) in which the electrical impulses form the information carriers. From the point of view of psychoanalysis, we are looking at the psychic apparatus in which information processing takes place. I have already explained that this approach requires the development of an overall

---

[1] Since SiMA currently runs simulations, the processing time is either extended or compressed, depending on the test requirements. Hence, the real-time requirements of the functions have to be taken into account in the experiments from the very beginning. I will have to go into this in more detail.

D. Dietrich, V. Hartmann Cardelle, *Simulating the Mind II*,
https://doi.org/10.1007/978-3-031-69530-8_6

model for both methods of description, in which the top-down principle must be observed. According to the extended Mealy principle, the top in our case is the psychic apparatus. This means that only after the structure of the psychic apparatus has been modeled can we move on to consider (and model in more detail) the underlying layers. Until then, these functions must and can only be specified in a rough form. However, the rough specifications are necessary to evaluate the model through simulation experiments, which always concern the behavior of the entire body. This means that the first important limitation in the following development is our focus on the psychic apparatus. The underlying layers are only addressed to the extent that they are necessary. Necessary means, first, that the information received from the body—i.e. from the neurological part—must be adequately processed or represented, and second, that the axiomatic and model concepts of the psychic apparatus must not contradict natural scientific principles and laws.

We know from information theory that the modeling of the lower layers only needs to be considered insofar as they have to meet the requirements that the psychic apparatus sets for the lower layers. This leads to the second limitation that must be strictly observed: As already mentioned in the previous chapter, the neuronal system of the Ψ-organ processes information asynchronously, i.e. each neuron processes its information independently of the other neurons and only depending on its synaptic inputs and its internal state. In the simulation program, in contrast to that, only a continuous flow of information with defined information cycles can be processed, in which feedback is of course taken into account. This requires an exact study of the real-time requirements, but this is not a general problem for control engineers. In all natural processes, and I would like to emphasize biochemical processes in particular, this is a well-known topic, yet in my experience the problem is often underestimated. Experiments—as in the SiMA project—are therefore the only adequate solution to validate the accuracy of the modeled results.

An additional note for those readers who are not yet familiar with real-time requirements: One of the purposes of a simulation program is also that the process time curve—i.e. the time during a simulation—can be stretched and compressed almost at will. This means that extreme cases, often up to physical limits, i.e. singularities, can be investigated with relative precision. However, the time cycles of the processes in the model must be in line with those of the simulation cycles. To put it more concretely: The digitalization of the functional sequences requires that the simulation cycles must run *significantly* faster than the actual internal mental process cycles—i.e. the sequences of the functions and the sequences in the functions—otherwise the simulation runtimes will enter into the behavior of the designed models. And this must not happen under any circumstances. I will discuss all these points in more detail below.

6.1 Goal of Modeling: What Makes People Human?

What exactly is to be simulated in SiMA? In AI and in the field of (engineering) cognitive science, the goal is usually to mimic human *behavior*. In contrast, the aim of SiMA is to understand *how* the Ψ-organ *works*. Why does the Ψ-organ—formulated more directly: and its individual *functions*, including those of the psychic apparatus—produce the *behavior* in question? We must therefore look for the *functional units* of the psychic apparatus that cause human *behavior*. The *behavior* itself can only point the way to an explanation, but cannot be the explanation itself. The explanation must be based on a *functional model*, a *structural model* of individual interlinked *functions*. Here again I would like to remind you of Valentin Braitenberg: Only after the functions of a vehicle with sensors, the motor, the control system, etc. are understood is it possible to explain how the process *vehicle* works. If the *functional model* is *simulated* in *experiments* and it displays a *behavior* that, within a certain degree of approximation, corresponds to the *actual behavior* of a physically existing vehicle, the developed model is considered evaluated and validated—which is the goal of such experiments.

The goal of this chapter with the topic of *model development* must therefore be to work out the *functions of the Ψ-organ* as well as their overall structure, which are then *simulated in experiments*. In natural scientific experiments, there is always the problem that it is not possible to start with complicated experiments if research in the relevant field is still in its infancy. This is the case with SiMA. We have to start with experiments that are easy to understand and, above all, easy to evaluate. Because if the functional structure generates a behavior that does not meet the expectations, the experimenter—in the case of SiMA the psychoanalyst—must have the chance to find the error. It must also be possible to break down all experiments into the simplest sub-processes that allow a clear, scientifically verifiable analysis and interpretation—without statistical methods. All this is the reason why the first experiments in SiMA are based on *people* (in SiMA we call them *agents*) who lived in a simple (manageable) environment around 8000 years ago. They are supposed to be about 30 years old, in good health and lively, without major physical or psychological damage. 30 years for two reasons. First, because psychoanalysis, pedagogy and psychology still know very little about children and especially small children. Thus, we are, for example, still unable to answer question why children learn so easily and quickly. Second, because degeneration processes occur in older people, which can have many different causes. The aim of the first experiments must be to test the relatively well established findings of psychoanalysis and neuroscience using scientific principles of simulation engineering. However, all of this first requires a first determination of *what constitutes the human in relation to other beings in nature*.

This question has so far been answered predominantly by philosophy.[2] For me, however, it is primarily a natural scientific question. I believe, I have found a

---

[2] For me as an agnostic, the subject of religion is not a question here.

plausible explanation for me. And that is a good thing, because I feel uncomfortable with philosophical answers. They are always influenced by their *zeitgeist*. They are based on education, schools of thought, subjective perceptions and logic. How often had philosophical insights to give way to natural scientific ones? With answers from the natural sciences, I think I am rather on the safe side. For example, I found excellent answers in the discussions between Henno Martin and Hermann Korn (Martin 2009, p. 132 ff, 180), who came to the conclusion that today's living beings are ultimately the result of a specialization that ensures their survival according to the Darwinian principle. Let us list just a few creatures: The cheetah is an extremely fast quadruped. The elephant is large and strong. The praying mantis has a fantastic ability to adapt to its environment. And the baboon has eyes that hardly miss anything. In the Darwinian sense, however, humans are far superior to all other creatures as individuals due to their intellect. Their intelligence, i.e. their mental abilities based on feelings, consciousness and language make them a unique creature that dominates all others, which they unfortunately also use extensively to eliminate others. So, *what distinguishes humans is the enormous capability of their Ψ-organs, the performance of their psychic apparatus, but certainly not their physique and muscles.* The development of mankind must be understood above all on the basis of the human Ψ-organ.

The neuroscientist Antonio R. Damasio elaborates this more precisely. He writes in Damasio (1994, p. XIII) that "... certain aspects of the process of emotion and feeling are indispensable for rationality ...". He substantiates this in detail with the findings of scientific, neurological studies of various cases. Affects, emotions and feelings are *valuation mechanisms* without which people would lose social contact with their environment (see also the book *The Spinoza Problem*—Yalom 2012), in which Irvin D. Yalom explains very clearly how crucial social bonds are for every human being).

Humans are characterized by their highly developed psychic apparatus with the following pillars, according to the findings of psychoanalysis,[3] on which the SiMA model is (initially) based:

1. The primary process (with unconscious information) is based on the pleasure-/unpleasure principle.
2. All perceived and stored patterns (objects, processes, ...) are valuated via a multi-layered system of quotas of affect, basic emotions up to feelings and are stored in a thus valuated form.
3. Stored patterns (objects, processes, ...) can be associated as memories.
4. Perceptions and activated memories cause conflicts in the primary process.
5. In order to resolve these conflicts, memories are sought that were sizes in the past with as much pleasure and as little unpleasure as possible and that are now associated with a corresponding quota of affect.

---

[3] Of course, the ten principles listed here are not all that psychoanalysis has developed to date. However, they are the ones to which we (initially) limited ourselves in the SiMA project, for a variety of reasons.

6. Based on this, a path to conflict solution is designed in the primary process that is associated from the currently perceived environment
7. A defense process in the primary process prevents the secondary process from being overwhelmed by a flood of possible solutions. This allows the secondary process to focus on essential activities as well as long-term preservation.
8. The primary process treats the information as *thing representatives*, the secondary process treats them as *word representatives*.
9. Consciousness is associated with word assignments to thing representatives. Consciousness therefore requires a learned language.
10. In the secondary process, the solutions that appear largely causally apt, realistic and logical are selected. The person becomes aware of the result—i.e. the solution that promises the greatest success.

There is no doubt that such criteria also apply to some animals, but the decisive point that sets humans apart is undoubtedly their enormous capacity for conscious awareness and rational thinking. But how do we learn to think rationally? Which mental functions are necessary for this? Which mental functions form the cornerstones of the Ψ-organ? Which (behavioral) abilities are part of the basic principles? I hope the next chapters will answer these questions. I cannot develop all the functions of the psychic apparatus, but must limit myself to the most important aspects of these functions. Because one thing is certain: The subsequently outlined development of the model of the Ψ-organ will remain a "never-ending story". And this story has just begun with Sigmund Freud.

I must refer you to Sect. 6.6 for a discussion of how we can imagine or even simulate consciousness. We have concrete ideas about this, but for reasons of capacity we have not yet been able to implement them—i.e. to model them in detail, let alone simulate them. However, in order to understand these particular considerations, the model of the Ψ-organ must first be understood.

6.2   Development of the Three-Layer Model of the Ψ-Organ

The goal is to simulate a functional model, i.e. a model consisting of interacting individual functions that as a whole generate human behavior. Valentin Braitenberg's experiments illustrate the relationship between the functions (such as sensor-, actuator-, electronic control-, and software functions) and the behavior of the vehicle. The functional model for this task can be designed on the basis of extended Mealy theory, which combines the descriptive principles of hardware and information engineering. This approach can also be used for modeling the Ψ-organ, because it is also an information system, albeit a highly complex one. For this goal, the necessary individual functions must be developed in their coherent structure. However, it must be taken into account that a functional model does not represent reality, but can only resemble it—as a scientist would say: *The model can only approximate (abstract) reality.* That is why it is called a *model*. It is crucial that the results of the subsequent experiments carried out afterwards show deviations,[4] we call this a certain Δ (delta). The maximum Δ between reality and the model that can be accepted must be determined *before* the experiments are carried out. If the Δ becomes too large in the experiments, changes must be made to the model and/or the experiments until the experiments produce results consistent with the underlying physical, chemical, psychological, or psychoanalytic theories, i.e. until the remaining Δ becomes acceptable. Excessive deviations can only be caused by faulty models, incorrect experiments, or programming errors.

Errors in functional models, and especially functional models that are wrong, will inevitably lead to unacceptable experimental results (think of the example of Nicolaus Copernicus in Sect. 4.3). So, there is no need to lock horns about incorrect model designs or to get upset about proposed solutions before doing the experiments. Because if the models or experiments are based on errors, the experiments will show behavior that does not correspond to the underlying theories anyway. Wrong models or wrong experiments inevitably lead, sooner or later, to irreconcilable contradictions in the theory: a basic principle of engineering. Of course, it is necessary to have a sufficient number of experiments.

The goal of SiMA, as we ultimately set it for ourselves, was to be able to evaluate which psychoanalytic theories are correct and which need to be modified (by psychoanalysts). Although the experiments so far have been simple in design and still need to become more varied and complex, and although we have not yet been able to conduct very many experiments, I can already anticipate that they will clearly indicate that the theory of psychoanalysis should in principle be correct. Overall, we researchers in the SiMA project are optimistic.

I have already mentioned why the project is based on psychoanalysis. As a brief reminder: in 1999, before working out the basic ideas for SiMA in concrete terms, I looked at the 17 psychotherapeutic schools in Vienna and discussed them with Heiner Bartuska, who had dealt with them professionally and officially (Bartuska et al. 2005). Only psychoanalysis considers the second topical model (*a functional model*) to be the decisive prerequisite for its discoveries. And after that, only psychoanalysis came into question for our research work.

Sigmund Freud distinguished three functions of the psychic apparatus: the id, the superego, and the ego. He first tried to link them to the neuronal model, but eventually gave up (Freud 2020f, pp. 282–284; Freud 2020a, p. 346). *With the extended theory of Georg H. Mealy, however, this step has become possible.* The first draft (Fig. 6.1) is therefore not very different from the sketches in Fig. 5.3, which I used to explain Mealy's theory.

The complete model thus contains the *neuronal system* (= neurological description) and the *psychic apparatus* (= information engineering description).[5] The complete model represents the Ψ-organ including the peripheral neurons and sensors

---

[4]Δ: Delta, fourth letter in the Greek alphabet.

[5]Psychoanalysts usually use the synonymous term mental apparatus instead of psychic apparatus. In the natural sciences, synonyms should be avoided as far as possible, since otherwise authors use

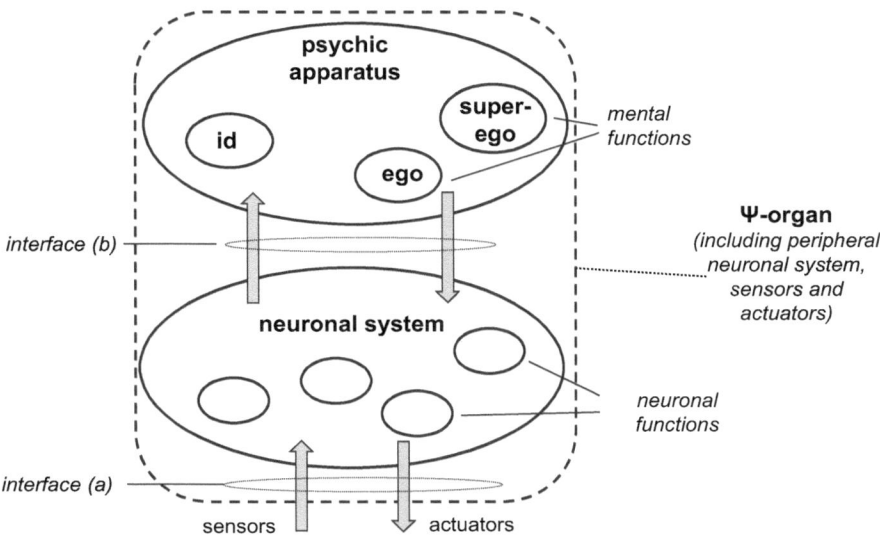

Fig. 6.1 All functional units with their integrated individual functions: *Ψ-organ* (= neuronal system, sensors, actuators and the psychic apparatus), *neuronal system* (= all neurons, including sensors and actuators) and the *psychic apparatus*

as well as the actuators (Fig. 6.1). In the lower box, the hardware is described on the basis of neurology, i.e. with physical including electrical, chemical and physiological, i.e. material-related methods and laws. According to Georg H. Mealy, the upper box, the psychic apparatus, is the information unit, which means that only information methods and laws can be used as a basis for description. The two lower interface components (the two arrows) of interface (a) represent input variables from bodily sensors (information inputs from eyes, ears, etc.) and output variables via actuators (information outputs via glands and muscles). The two interface components of the upper interface (b) between the lower and upper blocks represent the connection between the hardware models (the neuronal system) and the psychic apparatus. At first, I did not know how to model this interface. Only one thing was clear to me: The connection could not be trivial. I had to add a layer in between to be able to divide the huge functionality between the two layers into smaller functions. I had an image like Fig. 6.2a in mind. But what should this intermediate layer be? What could it look like? I needed help.

At the time I was developing these thoughts, I was on a research and teaching assignment in South Africa. My contact Gerhard Hancke (Professor of Engineering at the University of Pretoria) understood my problem and put me in touch with Mark Solms, a psychoanalyst and brain researcher working at the University of Cape

---

them with different definitions. In addition, it was important to give the system L2 + L3 (= mental apparatus) a name to simplify an efficient description. In SiMA L3 was named psychic apparatus.

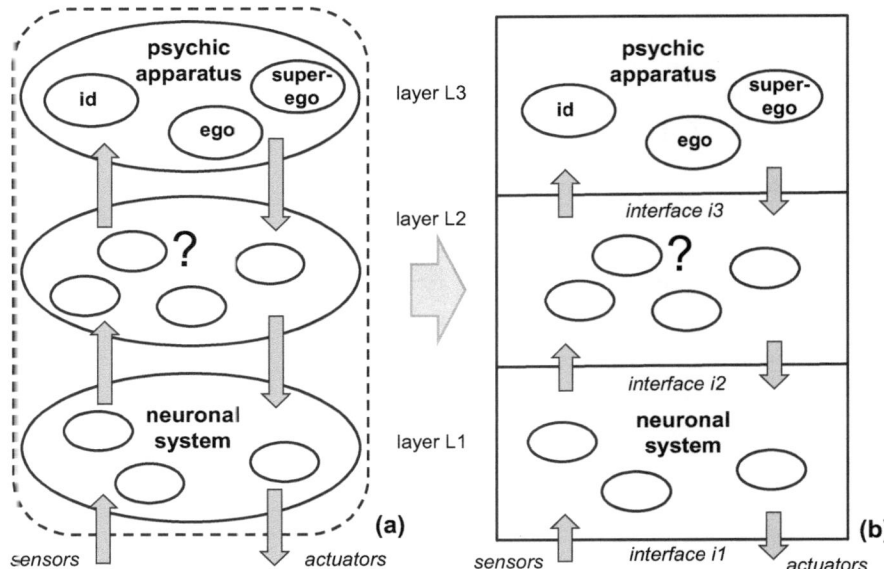

Fig. 6.2   First sketch of a functional three-layer model (**a**) *is derived from Fig. 6.1 by dividing the upper layer of Fig. 6.1 into two information layers;* (**b**) *is a simplified representation of* (**a**)

Town. This was the beginning of a long partnership (see e.g.: Dietrich et al. 2009, pp. 115 ff, 260). He listened attentively and—as he put it at a conference we organized together in Vienna in 2007—initially thought I was mad. How could an engineer take on such a challenge? I am neither a psychoanalyst nor a neurologist! The connection between the neuronal system and the psychic apparatus has long been a subject of study for many researchers. And now an engineer comes along and claims to have found the solution? We walked around his vineyard for over 2 h, and at the end he explained, as if it were obvious, that the second layer had to be that of symbolization. What did that mean? I had to let it sink in.

The task of interface i1 in Fig. 6.2b is clear. Neurology and other fields of medicine describe it in detail. The various human sensors transform different physical quantities (light quanta, sound waves, pressures, etc.) into electrical impulses, as I described in Chap. 3. However, which neurological functions are to be defined specifically in this lower layer (the nervous system) is of no interest in SiMA for the time being, since this depends primarily on which hardware is used as a basis. For the simulation experiments it is sufficient, based on Mealy's theory, that the lower layer fulfills the requirements of the higher layer. So, let us start with the i2 interface. The result of the pulse shaping (a pulse code modulation) of layer L1 is transferred to layer L2 via interface i2. There, through a complicated process called *neurosymbolization*, the patterns are formed with which the psychic apparatus can ultimately work. We can describe these patterns with symbols. The result of the processing in the psychic apparatus (layer L3) is sent back to layer L2 via the

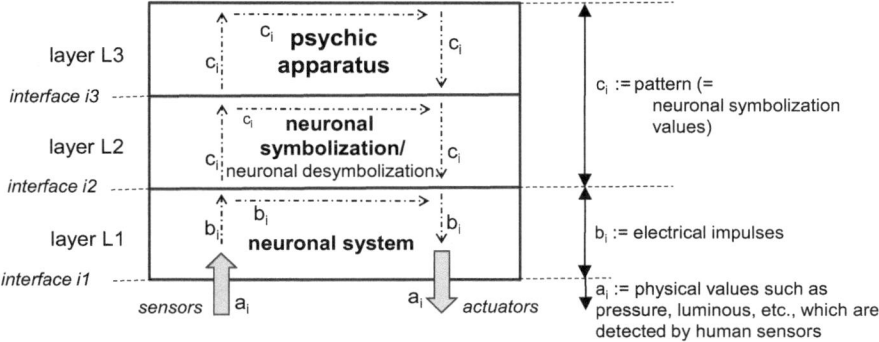

Fig. 6.3 Representation of the information flow (dashed arrows) in the model of the Ψ-organ and areas of the types of information description of a_i, b_i and c_i (For simplicity layer L2 is henceforth referred to as the layer of neuronal symbolization)

interface i3, which, after a process of *neurodesymbolization*,[6] sends the information back to layer L1. In layer L1, the processes appear to us again as electrical impulses, which ultimately stimulate the muscles and glands (= actuators).

So, the conversation with Mark Solms ultimately amounted to describing how the mental system, in our Fig. 6.2 the layers L2 and L3, processes, stores, associates, and forwards information, and finally how the interfaces i3 and i2 can be defined. Mark Solms made it clear to me that this is done on the same basis throughout the mental system (L2 + L3), regardless of where the information comes from, be it the ear, the nose, or the hands, or whether it goes back there. This concurs with the statements made by Jeff Hawkins (2004, pp. 63–64). In the SiMA project, we initially called these information variables *images*, regardless of whether they had an optical, acoustic, or other origin; today we speak of *patterns*. We describe them as symbolic variables that are obtained from the *electrical impulses* of layer L1 in layer L2 and, conversely, transformed back into impulses for the neurons of layer L1 as output variables.

Figure 6.3 shows the information flow graphically in more detail. Below the SiMA model, i.e. in the human body but outside of the nervous system (which is assigned to the hardware of layer L1 of the SiMA model), there are physical and chemical state variables that are measured via the sensors. In Fig. 6.3, the information a_i obtained there first travels upwards through layer L1 in the left-hand area as electrical impulses (signals; b_i) to layer L2 of the model, where symbolization takes place, i.e. the information is converted into symbols (patterns; c_i). These symbols c_i ascend to layer L3, the psychic apparatus, are processed there, then descend again to layer L2 (right in Fig. 6.3), and finally are transferred to layer L1 as electrical impulses b_i. As output variables of the nervous system, layer L1 activates the units of

---

[6]Neurosymbolization and neurodesymbolization are terms that are difficult to read, so for convenience I will use the terms symbolization and desymbolization instead.

the physiological body that are outside of the model of the Ψ-organ model (SiMA model): muscles and glands.

The horizontal arrows from left to right in layers 1 and 2 in Fig. 6.3 show direct feedback within the layers. In layer L1 they are called reflexes, such as the knee reflex (patellar tendon reflex) or the testicle reflex (cremaster reflex), which can run directly through inter-neurons in the spinal cord, allowing a human or other animal to react extremely quickly to stimuli without the information having to go through the psychic apparatus. I will discuss feedbacks within layer 2 in more detail in Sect. 6.3.6.

The thought process required to understand Fig. 6.3 is highly challenging. However, I have never claimed that the Mealy model is easy to understand. It is an abstraction with a mathematical background. In the Ψ-organ there is no real interface i2 at which anything is transformed. Similarly, there is no such interface between the hardware and the information unit in the computer, where electrical impulses are transformed into something else. The interfaces only serve the purpose of describing the different layers independently of each other. To help my students— they study computer technology and not computer science—to understand this enormous thought operation, I "forced" them in chip design to deal with concrete microprocessor examples and describe (which means to calculate) them in detail in various ways.[7] I wanted them to develop a "gut feeling" for functions in general. They should be able to describe and implement electronic units as hardware and software, i.e. approach the imaginary interface between hardware and software in different ways. Of course, the physical description (based on electrical engineering) and the corresponding realization in electrical circuits require an enormous effort and are computationally infeasible for large functions. In contrast, description and implementation in the form of software, i.e. the information engineering description of technical computers, is much simpler. I have repeatedly made the experience that the teaching method of having them apply both methods of description to the same function is what enables many students of electrical engineering (or computer engineering) to understand the extended Georg H. Mealy principle at all. Only then they start to comprehend the full scope of Georg H. Mealy's theory. I also maintain that those engineering collaborators in SiMA, who had subjected themselves to the "compulsion" of double, diversity calculations of different specific computer circuits,[8] were able to familiarize themselves more quickly with the subject matter of SiMA. In contrast to students of computer technology students of computer science generally do not have the opportunity to deal with this topic in this way.

---

[7] I always compare this process of learning chip design to learning to solve differential equations. You only get a "gut feeling" for differential equations if you have learned to calculate with them. If you only read explanations of differential equations, you have no chance of understanding them. In my opinion, this is probably also the reason why analysis and calculus can sometimes not be kept apart.

[8] I am thinking in particular of the design of machine code, which can be developed from the point of view of digital circuits or from the point of view of a machine language.

What is the most important conclusion to draw for the SiMA project? *If we want to simulate the Ψ-organ, we have to define an interface where we switch from one description method to another, i.e. from physical/chemical/physiological/etc. to informational engineering description and vice versa.* However, we must be aware that (by definition) we cannot find it in a real Ψ-organ. The Mealy principle is "only" a tool for the *description* and *modeling* of an information system.

The Ψ-organ is a good example of why pattern recognition is such an incredible challenge in engineering. Without question, the progress made in the field of robotics is remarkable, but humans are still far superior to robots in many ways. Experts have massive difficulties in designing a robot that can distinguish simple objects in nature with a high degree of accuracy. AI learned to work with symbolic values more than 50 years ago. However, modeling the functions that the Ψ-organ needs to derive symbols (patterns) from the electrical impulses of the neurons, with which the robot's Ψ-organ can then work as flexibly and with the same high quality as the psychic apparatus of humans, is still a major challenge for science (just imagine the task of developing autonomous off-road vehicles). Layer L2 (Fig. 6.3), the neuronal symbolization, is still an unexplored area of science. Our robots are far from having the capabilities of our human eyes, ears, and noses, not to mention the sensitivity of our skin or the adaptive control mechanisms of other organs. I will come back to this in a moment, but I will provide more detailed information on how to imagine the L2 layer model in Sect. 6.3.6. And one thing I am sure of: The L2 layer will keep scientists busy for a long time to come. There is no question that we have literally an *infinite* number of things to learn from nature.

Finally, I would like to mention three aspects that should make the next chapters easier to understand.

Aspect 1: The design of the model of the Ψ-organ according to the principle of Fig. 6.3 defines a crucial boundary condition: The division of the information layer into the layers L2 (neuro-symbolization) and L3 (psychic apparatus) according to the extended theory of Georg H. Mealy stipulates that these two layers are to be regarded as independent of each other. Only the condition of the underlying hierarchy has to be fulfilled: According to the top-down principle, the upper layer L3 must be designed first, and then layer L2 can be designed independently of layer L3, as long as layer L2 can fully meet all the requirements that layer L3 sets for it.[9]

When I say independent in this context, I mean that with the development of layer L3 (psychic apparatus), the interface i3 is defined between the two layers L3 and L2. This includes the requirements of layer L3 for layer L2. And on this basis, layer L2 can then be modeled in detail independently of the functions of layer L3.

Aspect 2: Aspect 1 contains a fundamental problem that leads to another aspect. I would like to change Fig. 6.3 to Fig. 6.4. According to psychoanalytic theory, the psychic apparatus (layer L3) is divided into a primary and a secondary process. The secondary process contains functions in which information is capable of

---

[9] The same applies to L1. Layer L1 can be designed independently of layer L2, as long as layer L1 can fully meet all the requirements that layer L2 sets for it.

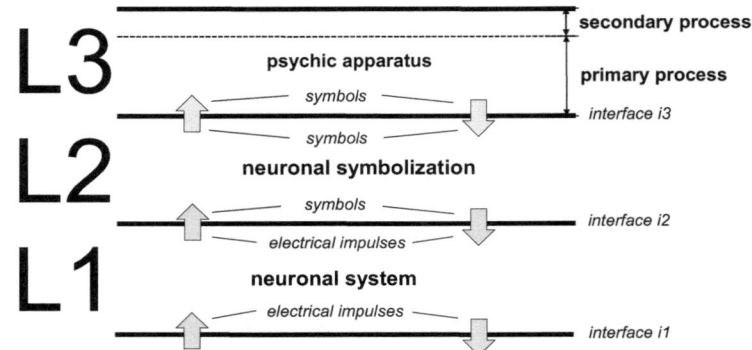

Fig. 6.4 Hierarchical functional layer-model of the Ψ-organ (with the layers L3, L2 and L1)

consciousness, i.e. can become conscious, as well as functions in which the information remains unconscious. In the primary process there are only functions in which information remains unconscious. For the psychoanalyst this means that the neuronal symbolization layer L2 can never be conceived through functions of the psychic apparatus. The primary process of L3 can only be observed indirectly.

If one wants to study the neuronal symbolization layer "from below", i.e. from the nervous system (layer L1), the neurologist is faced with the following problem: One has to structure the billions of neurons in such a way that the functions of layer L2 are crystallized and ordered from this enormous quantity. This is basically impossible.[10] The two layers above layer L1, the layers L2 and L3, are information layers and cannot be assigned to specific functions of the hardware layer (L1) according to Mealy's theory. This becomes clear with a little thought experiment.

Imagine inserting a needle into the central computer chip of your smartphone to extract impulses from the billions of internal wires. Suppose this was physically possible, but you did not have the schematics of the smartphone's apps, which were developed via software: How would you develop the schematics of the app's functions based on that? For me, it is simply inconceivable. We do not know the software of the Ψ-organ, i.e. how the design of the information system, and we do not have any plans for the Ψ-organ like we do for a smartphone. This means that layer L2 of the Ψ-organ is neither "accessible" from "above" nor from "below". For Mealy specialists, however, this is an almost daily problem during development. Often an additional layer has to be integrated between two layers, and you can only define the corresponding interfaces. This is exactly the case here. If the two

---

[10]Current imaging methods are far from sufficient to work out the functions of the neuronal symbolization layer L2 and the psychic apparatus (layer L3), since the resolution of today's devices is not capable of capturing the individual electrical impulse sequences of the billions of neurons and thus cannot be decoded (Hasler 2012; Huber 2012). The processes required for such decoding are not even remotely conceived, let alone understood. The challenge of developing the functions of layers L2 and L3 must be approached in a different way, and this is precisely the goal of SiMA, for which we have been able to develop a solution.

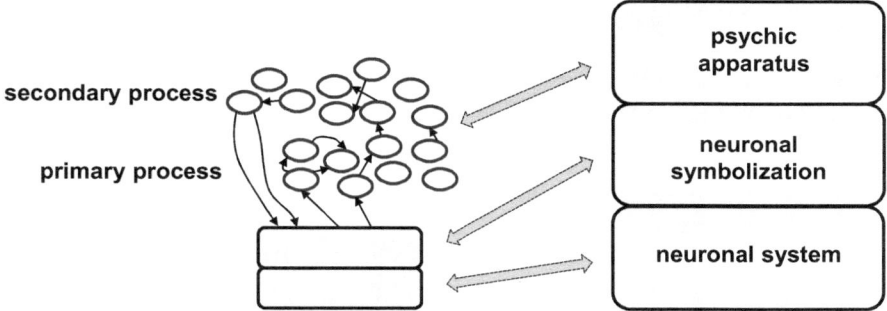

Fig. 6.5 Hierarchical three-layer model of the Ψ-organ with display of the distributed functional structure of the psychic apparatus

interfaces i2 and i3 are clearly defined according to Fig. 6.4, layer L2 can be developed based on these interface definitions. What remains to be addressed is the question of effort, and as mentioned above, I will address this in Sect. 6.3. Decisive for SiMA are only layer L3—the psychic apparatus—and, modified by it, the definition of the interface i3.

Aspect 3: The explanation in Sect. 3.3 may already allow us to see that the functions of the psychic apparatus are not hierarchically structured. I would therefore like to add a third aspect to Fig. 6.2, which leads me to Fig. 6.5. The elliptically depicted functions of the psychic apparatus in Fig. 6.2 and the flow of information between these functions are still arbitrarily designed here. However, the exemplary data flow and the approximate structure show that the psychic apparatus takes on an additional role. Although it still contains the hierarchical division between primary and secondary processes, the functions are otherwise heterogeneous, so that it can be regarded as a historically/psychically evolved structure. The relationships between the functions are determined solely by the necessary flow of information, i.e. the internal course of the sometimes parallel processes.

Figure 6.5 is also intended to convey something else: Each layer represents a model in its own right, which relies on its own description method (physical/ chemical/physiological/etc. or information engineering methods). And all layers in combination represent a complete model of the Ψ-organ.

One crucial function of the body has been mentioned several times, but has not yet been addressed: The information system of the body's hormones. It has not yet been included in the SiMA model, although it has a decisive influence on the behavior of the various processes in the human body. Yet, I see it as a parallel system to the neuronal system that works much more slowly. Hormones are released by endocrine glands or cell tissue and act as biochemical messengers in the human body. For example, in Sect. 3.1, I mentioned that the spike behavior (firing) of neurons is dependent on hormones and can therefore greatly change the behavior of a neuronal network. Or to give a more concrete example: The release of adrenaline increases heart rate, which increases blood pressure. In this sense, the hormonal system can be conceived as a parallel communication model to the model of the

Ψ-organ shown in Fig. 6.4, which is coupled to the neuronal system in different ways and thus interacts with it through different channels. However, it is necessary to think about this in more detail. It is a large, complex subject. But since it works relatively slowly in relation to the nervous system, and in the first stage of SiMA's development only the basic theory of psychoanalysis is to be understood and evaluated, the communication system of the hormones cannot and need not be included. There is no question that the integration of such a hormonal information model is inevitable in a further stage of development.

6.3   The Three-Layer Model in the Five Abstraction Levels (AL)

The development of the three-layer model of the Ψ-organ in the SiMA project was laborious, time-consuming and nerve-racking. I knew that the first steps, as always with such functional information model developments, would be the most challenging ones. The initial ideas and drafts led to blatant contradictions, which raised many questions and enormous doubts as to whether they could really offer suitable solutions. We discarded structures and made modifications until we were able to accept the overall design. The many dissertations (each took 5 years on average) and our even more numerous publications (many cited in Dietrich et al. 2015) show how hard we worked to develop this SiMA model. *The worst problem was that classical AI considerations based on behavioral models, algorithmic principles, or psychological ideas that were at odds with the findings of psychoanalysis or information theory kept creeping in.* This was bound to lead to contradictions in experiments.[11]

It is impossible to explain the development history of the model in detail. It is also impossible to derive every function and explain it in detail due to the sheer amount of material. It would also be boring for the reader. This is where I see the problem with our scientific report (version 3) (Dietrich et al. 2015), although we have tried not to write overly detailed explanations. For details, we generally refer to the corresponding 19 dissertations and the even more numerous master and bachelor theses that I was able to supervise with my colleagues.[12] Crucial aspects will be mentioned at various points below. In this chapter (i.e. Sect. 6.3 and its subordinated

---

[11] Once again: theories of different psychoanalytic schools that lead to contradictions must not be accepted from a natural scientific point of view, because otherwise these contradictions will show up in the experiment at the latest and then require corresponding changes in the functional model or the experiments. *But in SiMA, most of the problems were due to faulty axiomatics.* Universities are negligent in their teaching in this area, which has a massive impact on interdisciplinary projects like SiMA. And one more thing: when I talk about different theories of different schools, I am not referring to their clinical conceptions.

[12] It goes without saying that the scientific contributions (Dietrich et al. 2015), the dissertation, and the master and bachelor theses are freely accessible, and most of them were written in English, as is customary at the TU Wien (Technische Universität Wien).

subchapters) I would like to explain the structure of the three-layer model, the individual *functions* at the different abstraction levels, and the basic tasks of the functions. Of course, this also includes crucial aspects such as the general *flow of information* in the model, the influence of the *valuations* of the patterns (quotas of affect, emotions and feelings), the tasks of the drives, and also how the *self* is to be seen in relation to the *ego* of the individual, which is still completely new territory in natural science today. Finally, I would like to consider the large, relatively unknown *layer L2*—neuronal symbolization layer—and especially how it can be imagined and modeled for simulation. The *data model* used in SiMA will also be addressed, describing the *principle of data storage* up to memory. The last chapter deals with the old and still unsettled topic of neurology: localization. Alexander R. Luria's model will also be considered. I think that Georg H. Mealy's principle can also shape this discussion in a practical way.

6.3.1   L3 in AL5: Distinction from the Outside World

Based on information theory, I would like to use Figs. 6.4 and 6.5 as a basis for developing the model according to the top-down principle. For a relatively long time, it was unclear how many levels of abstraction should be defined for layer L3 (the psychic apparatus) in order to describe it effectively. After much experimentation and discussion, we decided to use *five levels of abstraction* in the project. This means that the layer L3 is described most abstractly on the top level, i.e. the fifth level of abstraction. Since interfaces are always a crucial feature in information systems, they should be defined directly at this abstraction level. On the one hand, this is the information that a human being takes in from the environment, and on the other hand, it is the information from their own body. I avoid the term *external world* here, because it is subject to a different definition in psychoanalysis, which I consider to be excellent. In psychoanalysis, the *external world* is everything outside the psychic apparatus (see Fig. 3.18), i.e. everything material (including the human physiological body and layer L1) as well as layer L2.

This means that we must distinguish between four types of input received by the human sensors: (1) Information from the world of states outside the human body, (2) information coming from the sensors of muscle states as well as the pain sensors, (3) information reflecting the states of homeostasis—think for example of the enteric nervous system—and (4) the information transmitted by the hormonal states. As outputs of the model, one could distinguish between those that control muscles and those that control endocrine glands. However, I have realized that this distinction does not play a role in the top-down development of layer L3. So, in the following I will consider only one common output for each layer L1 to L3. These considerations result in Fig. 6.6.

Each layer forms for itself a function that can be described at different levels of abstraction. Since I have defined five levels of abstraction for the psychic apparatus (see Fig. 5.10), Fig. 6.6 shows layer L3 at the fifth level of abstraction. This fifth

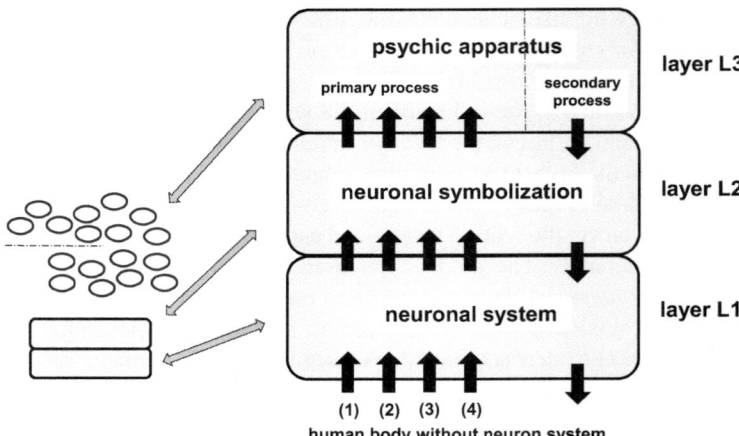

Fig. 6.6 SiMA model with the three functional layers in abstraction level 5 of layer L3 *(Information from (1) the world outside the human body, (2) from muscles and pain sensors, (3) from human homeostasis, and (4) hormones)*

level of abstraction must now be broken down step by step to level 1 according to the top-down design. Layers L2 and L1 are not considered in detail for the time being and are therefore excluded from further *detailed* development.

Figure 6.6 also illustrates something else: The coupling of the layers and their mutual dependencies can only be determined by the information to be described in the respective layers. The ascending and descending information, whose carriers are their data, are thus the link between the layers, a crucial dogma of Mealy's theory.

6.3.2   L3 in AL4: The Functions Id, Superego and Ego

Sigmund Freud, following Immanuel Kant, emphasized before the advent of the epistemological stance of radical constructivism that human subjective perception does not correspond to external reality (Freud 2020b, pp. 274ff). The psychic apparatus manipulates and valuates what is perceived before it becomes conscious. What functions are responsible for this?

Sigmund Freud outlined three functions for this purpose: the *id*, the *ego*, and the *superego*. They represent instances in the psychic apparatus that process incoming information. After various decision steps, they deliver their results to the muscles and glands of the person, whereby the person acts. In other words, the person behaves in some way. The model in Fig. 6.6 must be broken down accordingly for the next lower level of abstraction 4, which leads to Fig. 6.7.

The instance *id* is a function that determines the body's drive-wishes and receives the perceptual information from the neuronal symbolization layer L2 (in Fig. 6.7b the right model, left arrow coming from the neuronal symbolization layer which

Fig. 6.7 (a) Abstraction model: breaking down the function psychic apparatus on abstraction level AL 5 into its three main functions in abstraction layer AL 4; (b) functional model: the L3 layer, broken down to the abstraction level AL 4, with the three main functions and their information inputs and outputs

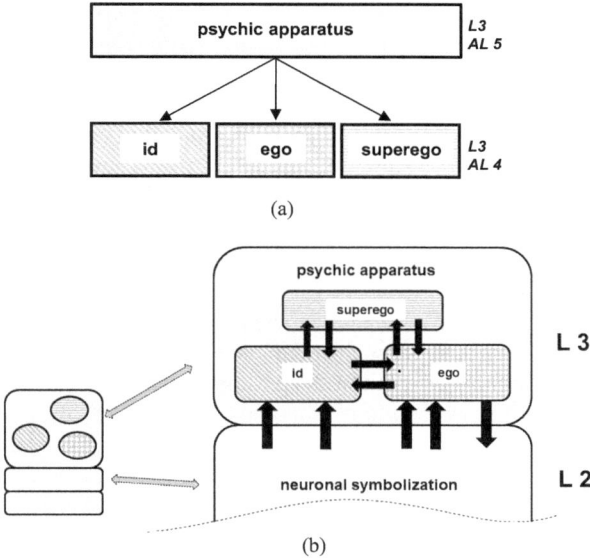

leads into the id). The drive-wishes are on the one hand the *drive representatives* and on the other hand *representatives of homeostasis*. The arrow to the right of it stands for the representatives generated by the muscular and perceptual systems of the external body world. The id operates according to the pleasure principle. The task of the id is therefore to achieve satisfaction of drive-wishes by seeking solutions in memories that promise success. As people store more and more memories as they grow older, there are always many, many solutions for every wish that demands satisfaction. Some of these are opposed by the superego, which is why the superego acts back on the id. The id offers its suggestions to the ego. The id is part of the primary process. Its contents (information) are therefore unconscious.

The next instance, the *superego*, is a function that includes prohibitions, commands and gratifications. It develops rapidly during infancy. However, this unconscious information, with which the superego operates, must not be confused with the social rules that can be made conscious, which belong to the functions of the action selection track of the secondary process (Fig. 6.11), which is explained in more detail in Sect. 6.3.5.

The ego must ultimately choose one of the many possible solutions to the conflicts that arise. For this reason, it is constantly in direct contact with the other two functions.

I think that the constellation of these three functions, which are called instances in psychoanalysis for a reason, illustrates well that (freely quoting) (Freud 2020c, p. 29): *A part of one's mental life escapes conscious knowledge and command of the ego's will*, or, as Sigmund Freud put it somewhat more theatrically, again freely quoting: *The ego is not master in its own house.*[13] Figures 6.6 and 6.7b certainly do not show the proportions to scale; the unconscious thought processes presumably occupy a much larger functional psychic area than the conscious ones. This consideration is consistent with the realization that the amount of unconsciously associated possible solutions far exceeds the amount of conscious information that the psychic apparatus has to process. Just think of the millions of sensors in a person that are constantly sending information to the Ψ-organ that, as perceptions, evoke associations. These have to be pre-processed and filtered in the primary process so that the person can focus their conscious thought processes on the event that they consider most important.

Figure 6.7b is intended to help to go a little deeper into the way of psychoanalytic thinking and to show the parallels with the way of thinking of computer engineering. Psychoanalysis looks at the psychic apparatus from three points of view: the *topical* point of view (how it is functionally structured), the *dynamic* point of view (how the contents interact in a conflictual manner), and the *economic* point of view (how the contents are valuated—in technical terms we would also say *weighted*—with psychic intensities (Freud 2020b, p. 282)). The topical point of view becomes particularly clear in Fig. 6.7b. In Sigmund Freud's *second topical model* (Freud 2020f, 2021b) three functions were defined. Psychoanalysis often even uses the term *entity* for this, a term that has been introduced into digital communication technology in the same sense.

The *first topical model* distinguishes between the unconscious, the preconscious and the conscious (Freud 2017, pp. 407ff), which of course contradicts the second model.[14] In the natural sciences, two functional structures cannot contradict each other. Psychoanalysis has not yet been able to resolve this contradiction. However, once one becomes familiar with the extended theory of Georg H. Mealy, the solution becomes apparent. I have already mentioned this consideration. If we view the terms unconscious, preconscious and conscious as kind of *behavior* of information, i.e. *properties of information*, this first model is, in the sense of computer

---

[13] Sigmund Freud uses a very clear metaphor (Freud 2020f, p. 283). The *ego*, an instance (we call it a *function* in SiMA), is just a small rider sitting on a big horse, the *id*. The question is always whether the little rider can keep the big horse steady. Whereas the horse, which must not be forgotten, always has a life of its own. For Sigmund Freud, however, the crucial difference between metaphor and reality is that a real rider has forces that are independent of those of the horse while the *ego* obtains its psychic intensity, from the *id*—how this comes about, I will explain later. *Without psychic intensity, the human being would be an extremely inflexible creature.*

[14] Sigmund Freud himself recognized this problem and found himself unable to come up with a solution (Freud 2021b, p. 385) (freely quoting): "W*hat is the actual nature of the state that reveals itself in the id through the qualities of the unconscious, in the ego through those of the preconscious, and what is the difference between the two? Well, we know nothing about this, and our meagre insights stand out miserably enough against the profound darkness of this ignorance. Here we have approached the actual as yet unrevealed secret of the psychic.*"

engineering, a *behavioral model* and not a *functional model*.[15] And please note: A functional model (like the second topical model) and a behavioral model not only complement each other, but presuppose each other. This is because the premise of Georg H. Mealy's theory is that *functions are the generators of a process, i.e. they exhibit a corresponding behavior*. There must therefore be two models, one for the structure of functions, the other for the behavior of functions. *Sigmund Freud thus developed parts of the foundations of information theory that computer engineering and information engineering in general did not reach until the 1960s.* However, computer engineering deserves the honor of being the first to have worked out a clear distinction between functions and their behavior in a natural scientific way.

Let us now turn to the second psychoanalytical point of view mentioned above. Consider Fig. 6.7b. The *dynamic* point of view of psychoanalysis states that in the constellation of ego, id and superego described above, conflicts are bound to occur. The whole system is in constant flux. The system can never be balanced or even remain in a static state due to the constant influx of information from the various external and internal states of the body. The changing homeostasis, the drive representatives and the contents of the superego constantly produce contradictory conditions that require solutions.[16] However, this also means that the ego would never be able to cope with this flood of primarily associated content if it had not developed powerful defense mechanisms (a special sub-function of the ego). The defenses must be consistently active at all times, even during sleep.

The third point of view, the *economic* one, is probably the most difficult to understand and to describe. In psychoanalysis, the economic point of view is referred to as the *cathexis* of psychic content by *psychic energy*, a choice of terminology that today leads to serious misinterpretations. The concept of energy is today firmly anchored in physics and technology worldwide and is also legally standardized (nationally and internationally). Since Albert Einstein, at the latest, the concepts of energy and mass can no longer be viewed independently of each other.[17] Today they are clearly an integral part of the conceptual world of physics and only of physics. This is how they are taught in our schools. In the back of our minds, the term energy is always associated with mass, with something that flows. Yet, *in our psychic apparatus we live in the world of information.* Adopting physical definitions or even using metaphors from physics is dangerous. I have pointed this out many times. They come from a different world and are inevitably subject to different methods and

---

[15] So, as long as one speaks of *the* unconscious and *the* conscious, one sees the first topical model as a functional model and that means it contradicts the second topical model, since in natural sciences you cannot have two models saying two different things.

[16] Remember, according to psychoanalysis, observable behavior can be the result of a conflictual interaction between the perceived information and the psychic content of the functions.

[17] It must be remembered that the term energy was not used axiomatically until Albert Einstein's theories defined it. Before that, it was a term for many things. Due to the national and international legal definition of the term, as well as its adoption by schools and universities in their teaching, it is necessary to question other earlier meanings—such as psychic energy—and possibly give them a different name, so as not to generate false associations, especially of its properties.

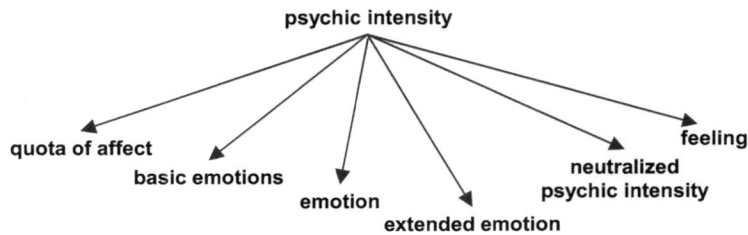

Fig. 6.8 The valuation variable psychic intensity (referred to as psychic energy in psychoanalysis), broken down into the valuation variables defined in SiMA, which operate in different functional areas of the psychic apparatus

laws. And in this case, information engineering, they certainly are. *In the psychic apparatus—i.e. in the world of information—nothing physical flows,*[18] also no physical energy that could possibly be "recharged", as some people naively imagine. Nor can it be held in containers. This is why the term *psychic intensity* had to be introduced in SiMA. Psychic intensity is assigned to all psychic content—i.e. all objects, patterns, representatives, etc.—by which they are valuated—as psychoanalysts say: *cathected. Psychic Intensity* is in principle timeless and massless.

Psychoanalysis sees the source of the principle of valuation in the hardware, in the neurological (layer L1), in the homeostatic tension (hunger, thirst, etc.), where actual matter flows. Sigmund Freud speaks of the *sexual tension* (my translation from "Sexualspannung" in Freud 2018a, p. 69). They are recorded via body sensors and symbolized (transformed) in the neuronal symbolization layer L2, into drive representatives with which the psychic apparatus (layer L3) can work. They are therefore only information and cannot be physically described.

However, the rather general psychoanalytic concept of *valuation* could not be maintained during the development of the SiMA model. An axiomatically clear and sufficiently differentiated definition of the term had to be developed. The result is shown in Fig. 6.8 and is explained in detail below. It is based primarily on the considerations of Antonio R. Damasio (1999, pp. 35ff). The individual terms do not contradict the psychoanalytic way of thinking, but rather represent an extension of it. Antonio R. Damasio distinguishes between different emotions and the feelings. They are assigned (mathematically in SiMA) to different areas of the psychic apparatus and are described vectorially. On the one hand they characterize the SiMA model, on the other hand they show how difficult it is to reproduce the human psychic apparatus as accurately as possible.

This makes it clear that the concept of *psychic intensity* serves as an umbrella term for all valuation variables.

---

[18] According to the theory of psychoanalysis, in the primary process cathexis (valuations, weightings) can be *displaced* from one content to another. Even in such a case, nothing *flows physically*, only the *information of valuation* is *transferred* from one content to another.

The most hardware-oriented *valuation variable* of the psychic apparatus is the *quota of affect*. The quotas of affect are based on drive representatives. They are fed directly from the neuronal symbolization layer and are input variables of the psychic apparatus (inputs into the id in Fig. 6.7b). They operate purely on the pleasure principle and not on the unpleasure principle, i.e. they follow the logic described by the question: Which drive object or drive aim promises the highest pleasure? The reality principle, on the other hand, does not play a role in the primary process, which assigns quotas of affect to objects of drive and aims of drive.

Emotions are determined by bodily and psychic states, and thus represent the state of a person from the perspective of the primary process. They are not only valuated according to the *pleasure principle*, but also according to the *unpleasure principle*, i.e. they also try to avoid *unpleasure*. This means that pleasure and unpleasure can occur independently of each other, i.e. high pleasure can be accompanied by high unpleasure.[19] *Emotions* are calculated from *current quotas of affect*, *current pleasure* and *unpleasure*, and valuated *memories* associated with *perceptions* and *fantasies*. This also means that the amount of affect can be seen in the emotions. This is important for the psychoanalytic principle of the *mobility* of cathexis (valuation), because the size of all quotas of affect remains the same in all individual valuations.[20] Valuations can be condensed and displaced in the primary process, but not in the secondary process (Freud 2020c, p. 421). The displacements occur through chains of associations. This can lead to a reduction or increase in psychic intensity.[21] The result are contexts of meaning that are often not easy to explain at first glance. In dreams this phenomenon becomes apparent.

This also means that during the transition from the primary to the secondary process, freely "mobile" (better: displaceable) psychic intensities are firmly bound to specific presentations.

---

[19] Just think of sadomasochism.

[20] Freely quoting from Breuer (Breuer and Freud 2015, p. 310) who in this quote refers to Freud: "The fully regenerated brain elements therefore release a certain amount of energy even at rest, which, although not functionally utilized, increases intracerebral excitation. This creates a feeling of unpleasure. Such feelings arise whenever a need of the organism is not satisfied. Since the ones discussed here disappear when the excess of excitation released is functionally utilized, we conclude that this elimination of excess excitation is a need of the organism. Here we encounter for the first time the fact that in the organism there is a 'tendency to maintain intracerebral excitation at a constant level' (Freud)". And later in his career Freud wrote (Freud 2020c, p. 399) (freely quoting): "The psychic apparatus strives to keep the amount of excitation that is present in it as low as possible, or at least to keep it at a constant level."

[21] Psychoanalysis speaks of *discharge* or *accumulation*, terms that we should avoid in SiMA for axiomatic reasons, as they are associated with physical laws and indicate something that flows. However, information does not flow in the sense of information engineering. The term "flowing" could axiomatically only be used for its physically describable carriers. In common parlance, however, the idiom is used: *"This or that information flows into this or that project."* Such a formulation is a showcase how common parlance often is, by the standards of a precise physical and computer engineering description, a poor formulation that expresses a lax attitude, which I sometimes (guiltily) follow as well.

Characteristic emotions that are passed all the way through the functions of the psychic apparatus until reaching its outputs are the *basic* and *extended emotions*. Therefore, they also determine the behavior of the body. As the name suggests, the *basic emotions* comprise are a few specific emotions. Six are considered in SiMA (Dietrich et al. 2015, p. 71) (see also Fig. 6.15 below):

- Joy
- Anger
- Anxiety
- Mourning
- Love-saturation
- Love-elation

We have introduced the term *extended emotions* in SiMA in order to adequately describe emotions that are the result of defense mechanisms—which, of course, are based on basic emotions—and that are indicative of certain behaviors. In SiMA we distinguish between (for further explanations see: Dietrich et al. 2015, p. 71; Deutsch 2011, p. 59):

- Envy
- Greed
- Pride
- Pity
- Guilt
- Depressive mourning
- Shame
- Disgust
- Hate with object
- Love with object

Their formation is the result of complex processes that cannot be discussed in detail here. For those interested, I refer to Dietrich et al. (2015), Schaat (2016), List (2009, pp. 81, 91ff).

While emotions, according to Antonio R. Damasio, as mentioned above, valuate unconscious information (patterns representing objects or aims) in the primary process, feelings act in the secondary process. They are calculated from the *vectors*[22] *of emotions* and are described as *scalars*. Depending on the function in the secondary process, they occur unconsciously, preconsciously, or consciously. A crucial feature of feelings in contrast to emotions is that they are connected to *word representatives*.

If we look at the whole structure of the differentiated *valuation* of patterns in the psychic apparatus, it becomes clear that nature has developed a sophisticated principle that is still far superior to all technical systems of AI in terms of

---

[22] A vector is composed of an arbitrary number of scalars. A scalar is a quantity determined by a numerical value.

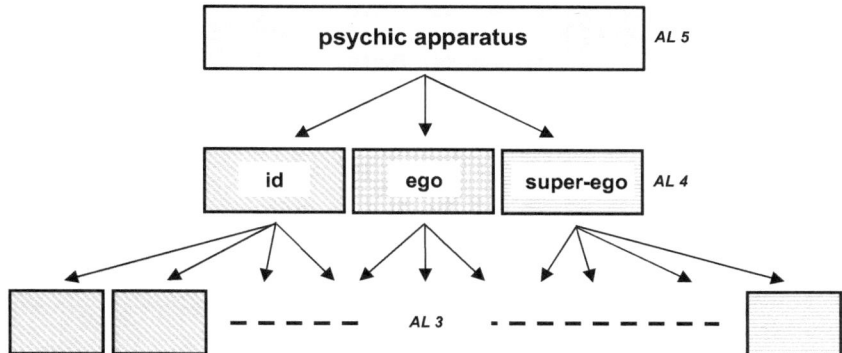

Fig. 6.9 Starting from Fig. 6.7a, layer 3 is broken further down to abstraction level 3 (AL3)

flexibility.[23] The ability of the primary process to make associations alone is enormous—which is why I spoke of brainstorming in Sect. 3.3—but it remains unconscious and does usually not interfere with the secondary process, it only provides it with filtered content. People's consciousness is therefore focused on what they consider important. They are therefore much better prepared than other living beings to deal with new situations. This is a crucial aspect that needs to be systematically incorporated into AI.

6.3.3   L3 in AL3: The Functionally Structured Mental Tracks

Following the top-down development, layer L3 is further broken down to the next level of abstraction—abstraction level 3 (AL3)—which is depicted in the *abstraction model* in Fig. 6.9. In order to keep the entire *functional model* in mind, layers L2 and L1 are included in the figure with a few functions each. However, the previous vertical representation is difficult to display graphically, so the model is tilted by 90° (Fig. 6.10). All this leads to the final, basic representation in Fig. 6.11.

Figure 6.12 shows the progression of information through the psychic apparatus in a simpler way. Let us take a closer look at the tracks for the general progression of information.

The information progresses through these functions one after the other, so in SiMA we came up with the idea of calling them tracks. The information is manipulated on these tracks. They develop quotas of affects, emotions and feelings and are linked to associations. Thing and word representatives are never stored without a

---

[23]However, its disadvantages should not be ignored. It is often difficult to judge the simplest relationships in a reasonably objective way, because the valuation mechanisms are not easy to understand. Think of the accounts of Yuval N. Harari, who overturns previous historical conceptions and thus can answer some questions (Harari 2015).

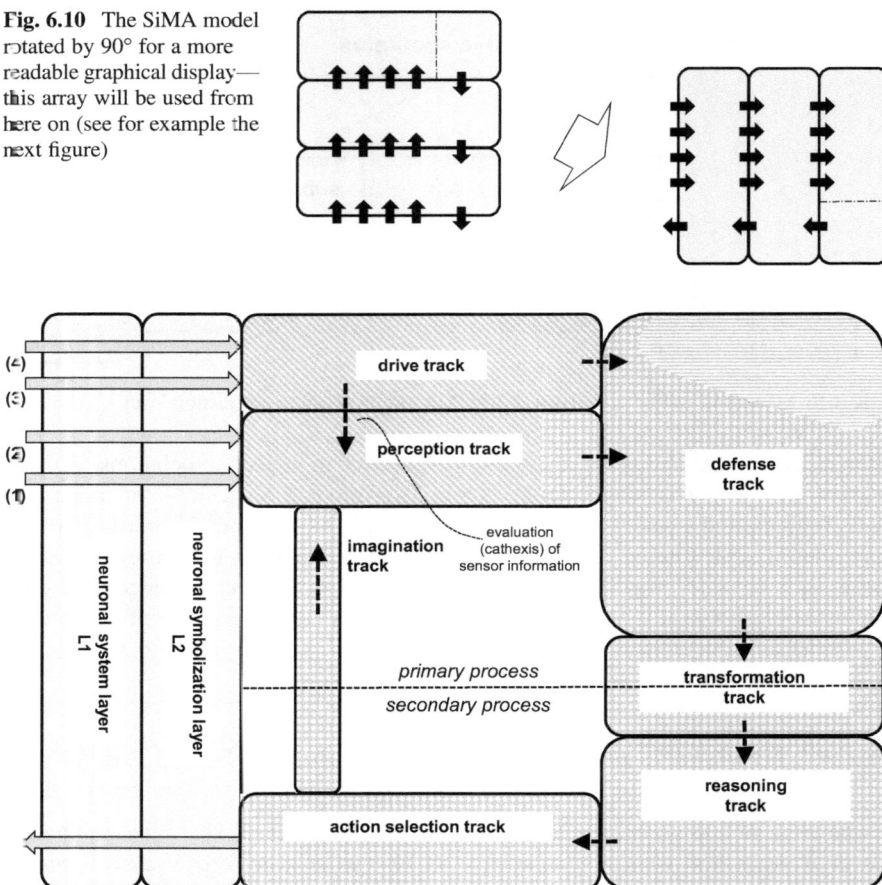

Fig. 6.10 The SiMA model rotated by 90° for a more readable graphical display— this array will be used from here on (see for example the next figure)

Fig. 6.11 SiMA model in abstraction level 3 (AL 3) (1) environment perception track, (2) body perception track, (3) self-preservation drive track, (4) sexual drive track

valuation, or conversely, associated *thing* and *word representatives* are always *valuated* (psychoanalytically formulated: *cathected*). The wish to be able to associate something neutrally or objectively is very difficult to fulfill consciously, unconsciously not at all. An example: my favorite cup from Fig. 3.17 is strongly positively valued, other cups less so, but they all have a meaning in my memory (psychoanalytically formulated: the cups are differently cathected).

According to the four inputs listed in Fig. 6.6, we must distinguish between four different input tracks in layers L1 and L2 of the model shown in Fig. 6.11: the (1) *environment perception track*, the (2) *body perception track*, the (3) *self-preservation drive track* and the (4) *sexual drive track*, which form the input of the psychic apparatus. The actuator track is the output.

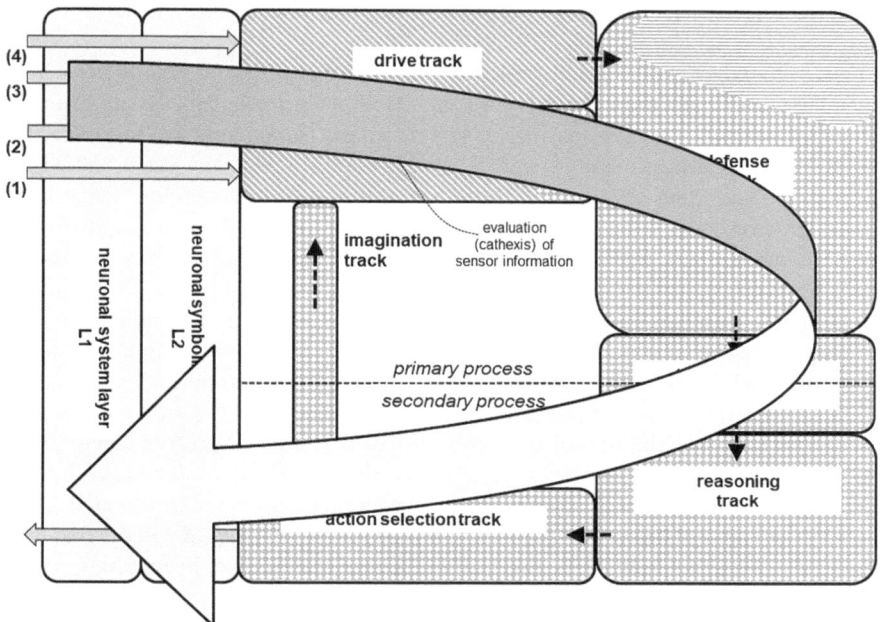

Fig. 6.12 Progression of information in the SiMA model in the third abstraction level

The information of the sexual drive track (drive representatives and quota of affects) and the self-preservation drive track (thing representatives of homeostasis) is determined in the function of the *drive track*. The information of the body perception track (muscles, pain, temperature, etc.) and the environment perception track (visual, acoustic, olfactory information, etc.) flow into the *perception track*. According to Sigmund Freud (2020b, p. 269), the information from both tracks is not conscious (see also (Sacks 1985, p. 67).

The information in the drive track influences the perception track (dashed arrow in Fig. 6.11). When you are hungry, your perception of food is stronger and you buy more food than when you are not hungry. If you have elevated levels of sexual drive hormones, you perceive the outside world differently than you normally would. Interesting, but understandable, is the fact that information from the perception track does not affect the drive track. The drive needs are independent of the person's environment.

The manipulated information from the drive and perception track reaches the *defense track*. Any conflicts that arise are attempted to be held back with various defense mechanisms. The information that can still pass through the defenses, i.e. that is to be transferred to the secondary process, must be made *capable of becoming conscious*. This is what the *transformation track* is for. The secondary process is made up of the *reasoning track* and the *action selection track*, whose output information ultimately controls the muscles and glands via layer L2 and layer L1.

The secondary process is governed by the reality principle. It is only here, in the reasoning track, that the proposals for action from the primary process are examined to check whether they are feasible or just a fantasy game. Here the following gets worked out: Where do I really stand? Where do I need to focus my attention? What is my short- and long-term goal? There are always several solution scenarios to consider, from which the action selection track then selects the solution that promises the most pleasure and the least unpleasure, taking into account certain factors such as external social rules. Finally, the action selection track also has the task of sending the right patterns to the neuronal symbolization layer to stimulate the many different muscles and glands.

The *imagination track* is a feedback track. The unused solution scenarios are not eliminated, but fed back into the perception track of the primary process as *fantasies*. They still influence the further actions of the psychic apparatus. There are certainly more feedbacks in the psychic apparatus that will have to be worked out later. In principle, they represent a major problem that should not be underestimated, which is why they are not taken into account in SiMA for the time being. On the one hand, feedbacks are absolutely necessary.[24] There are several of them in every layer of the model, just think of the reflex arcs in layer L1 that I mentioned earlier. On the other hand, it is important to know that it is very easy for the process to become unstable if the feedbacks are poorly set. The instability depends on different parameters of individual functions. And when simulating a process, it is important how fast the simulation process runs in comparison with the real psychic process. This requires the experience of the experts who develop the simulation, but this is not to be discussed here. I must refer to the reports on the results of our projects, and especially to the dissertations on which the SiMA project is based.

6.3.4   L3 in AL2: The Drive Aspects in More Detail

In abstraction level 2 (AL2) of the psychic apparatus (layer L3), the primary process in particular should be considered in more detail (Fig. 6.13). It can be seen that the *sexual-* and *self-preservation drive tracks* of layers L1 and L2 are directly carried on in layer L3.

The body develops drives, which, according to (Freud 2020a, pp. 234–235), develop into drive energies (i.e. drive intensities). For our model, this means that the drives physiologically recorded in layer L1 are symbolized in layer L2. This

---

[24] Some readers may be interested in the article (Dietrich 1984), which explains how feedback gained increasing importance—in the sense of information engineering—in automation and especially in 1984 in the Airbus A320. And not only for control-related reasons, but also for safety reasons. Jeff Hawkins also deals with this topic in detail (Hawkins 2004, pp. 88–89, 163).

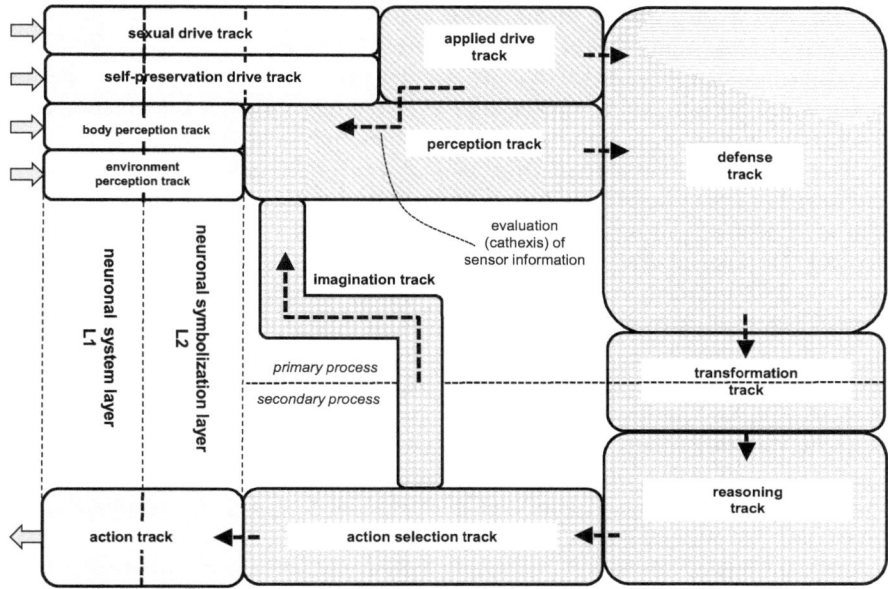

Fig. 6.13 The SiMA model in the layer of abstraction 2 (AL2)

generates, first, the *quota of affect* in the sexual drive track, which represents the quantity of *drive tension* and the first valuation variable in the psychic apparatus, and, second, the following psychic contents: the *source of drive representative*, the *aim of drive representative* and the *object of drive representative*. The source of drive representative is the representation of the internal somatic source of a drive. According to its location, there is a phallic (phallus), an anal (anus), an oral (mouth), and a genital (concerns the entire sexual organ) representative. The *aim of drive* is an action through which satisfaction can be achieved, which requires a *drive object* in order to achieve the aim of drive. The partial drive representatives thus place the demand for a solution on the psychic apparatus. *It is interesting to note that the object of drive (be it a person, an object or a fantasized object) is the most variable of a drive representative*. It can be exchanged easily. The unit formed by the quota of affect and the drive representatives is divided into a *libidinal* and an *aggressive* component, which form the input variables of the *applied drive track*. The ratio of the division between the libidinal and aggressive components is taken as a *personality parameter*[25] in SiMA. The experts in our team have not been able to clarify to what extent this is genetic or acquired.

---

[25]Personality parameters are characteristic values of functions of layer L3 and thus characterize an individual. Each personality parameter of a function must therefore be determined or defined individually.

The drive representative of the *self-preservation drive track* reflects the *drive tension* of a body organ caused by an unbalanced homeostasis (e.g. hunger or thirst). The psychic apparatus has the task of solving the tension problem (reducing the homeostatic tension), but this (the drive representative of the self-preservation drive track) only leads to a wish for pleasure gain in the secondary process.

The main task of the *applied drive track* is to find solutions for the satisfaction of drive-wishes from valuated memories. In doing so, it takes into account drive-wishes that are under primal repression, and it processes and revaluates repressed drive content. The applied drive track has an additional difficult task, that of *desexualization*, in SiMA called *neutralization of intensity* (psychoanalysis speaks of the neutralization of drive energy). In the course of human development, the psychic apparatus becomes increasingly capable of discharging *psychic intensity* (a valuation variable) from drive representatives and making it available to higher cognitive functions of the secondary process. The degree of this ability reflects a person's maturity. In other words, as a person matures, they become increasingly able to obtain pleasure gains from cognitive processes. Only the neutralized intensity transferred to the secondary process enables a person to engage in complex thought processes.[26] In addition to the secondary process, the *defense function* is the winner of this transference of intensity. It requires a great amount of neutralized intensity in order to be able to change and repress psychic content. This is *one* of the reasons why strong neuroses can lead to a dramatic reduction in cognitive performance.

The remaining tracks shown in Fig. 6.13, from the defense track to the action track, do not need to be discussed in detail here. They are sufficiently covered in the last chapter and will be discussed in detail in the description of abstraction level 1 of the L3 layer. However, I think it is important to mention two other points.

The first point concerns the axiomatic terminology used in SiMA. In SiMA, a strict distinction is made between *perception* and *recognition*. *Perception* occurs exclusively in the primary process in the psychic apparatus and means that objects or actions are perceived unconsciously. They generate associations and evoke memories, i.e. they establish connections in the network of object representatives that are valuated with quotas of affect and emotions. *Recognition*, on the other hand, is much more: a process within the secondary process. It is a conscious process and establishes a relationship between what is remembered and what is perceived. Recognition is valuated emotionally.

The second point is intended to illustrate that the SiMA model increasingly raises questions if one goes into more detail.

The term *instinct* has deliberately not been used so far, as it is still very controversial in the scientific community (Erikson 2018, p. 15). Various scientists do not assign instinct to the psychic apparatus (Jacobson 1998, p. 63). Accordingly, it

---

[26] Sigmund Freud writes (Freud 2020f, pp. 299–300) (freely quoting): "If this displaceable energy is desexualised libido, then it may also be called sublimated. . . . If we include thought processes in the broader sense among these displacements, then thought work is also driven by the sublimation of erotic drive."

originates from a more ancient part of the Ψ-organ (Panksepp 1998, p. 44, 122; Solms and Turnbull 2002, p. 29). In the SiMA model it can therefore be assigned to the neuronal symbolization layer L2 or the neuronal layer, i.e. layer L1. In mammals such as humans, there are Ψ-organ functions that are genetically determined to trigger, for example, a search, anger, or fear response under certain conditions. We can call this an innate mechanism.

What does Sigmund Freud have to say about this? Let us just look at the seeking function. According to his understanding, drive representatives generate aims of drive representatives which represent a demand on the psychic apparatus to perform work. If there is a memory of having searched for something and found it, this memory represents a possible solution. This is how Sigmund Freud explains to me seeking via functional units of the psychic apparatus, i.e. layer L3. André Green explains it in a similar way. He also refers to unconscious processes of the psychic apparatus (Green 1975, p. 517; Dornes 2001, p. 172). If a subject, i.e. a human being or a highly developed mammal, has the impression that an object it once possessed has been lost, it goes in search of it. The very thought of being able to find it again promises a solution to its conflict and thus a pleasure gain. André Green assigns this process to the primary process because he classifies it as unconscious.

Jaak Panksepp, in contrast, sees searching as an instinct (Panksepp 1998, pp. 144–163), as a function, which is located under layer L3 and therefore part of layer L2 or even layer L1. However, the two theories mentioned above are not mutually exclusive. It is possible that several principles with different imprints are present in humans at the same time. The question is, how can we locate the search function in the Ψ-organ and describe its behavior sufficiently well? My answer is the same as always in such cases: Verify this by simulation experiments based on the SiMA model. This would be the task of the relevant experts, i.e. neurologists, behavioral scientists, psychologists and psychoanalysts. They should construct case studies, and the simulation must then show what is or is not correct by simulating and thus testing all possibilities.

6.3.5   L3 in AL1: The Individual Functions of SiMA

Abstraction level 1 of layer L3 (psychic apparatus) is subject to the lowest abstraction, i.e. it is the lowest level of abstraction. All the functions it contains are listed in Table 6.1. Figure 6.14 shows the graphical representation of L3 at abstraction level 1. This representation of the overall structure of the psychic apparatus makes it clear how the, earlier discussed, flow of information progresses through the individual functions of the model.

Further details of these functions of the lowest level of abstraction and their individual behavior can be described by working out the tasks of the individual

Table 6.1 Designations of SiMA functions in abstraction level AL2 (= tracks) and abstraction level AL1

AL2	Abstraction level 2 (AL2)	AL1	AL1	Abstraction level 1 (AL1)	Layer affiliation	Part of the psychic apparatus function
ST	Sexual drive track	F39	SeSy	Seeking system (libido source)	Layer 1 (neuronal layer)	
		F40	NeLi	Neurosymbolization of libido	Layer 2 (neuronal symbolization layer)	
		F64	QaSe	Generation of quota of affect for sexual drives	Layer 3 (psychic apparatus)	Id
SeT	Self-preservation drive track	F1	SeMe	Sensors metabolism	Layer 1 (neuronal layer)	
		F2	NeNI	Neurosymbolization of metabolism information	Layer 2 (neuronal symbolization layer)	
		F65	GeST	Generation of quota of affect for self-preservation drives	Layer 3 (psychic apparatus)	Id
aDT	Applied drive track	F48	GoDC	Generation of drive components	Layer 3 (psychic apparatus)	Id
		F57	GoDR	Generation of drive representatives	Layer 3 (psychic apparatus)	Id
		F49	pRDW	Primal repression for drive-wishes	Layer 3 (psychic apparatus)	Id
		F54	EbDC	Emersion of blocked drive content	Layer 3 (psychic apparatus)	Id
		F56	DeNe	Desexualization/ neutralization	Layer 3 (psychic apparatus)	Ego
BPT	Body perception track	F10	SefE	Sensors for the environment	Layer 1 (neuronal layer)	
		F11	NfEn	Neurosymbolization for the environment	Layer 2 (neuronal symbolization layer)	
EPT	Environment perception track	F12	SefB	Sensors for the body	Layer 1 (neuronal layer)	
		F13	NeBI	Neurosymbolization for body information	Layer 2 (neuronal symbolization layer)	
PT	Perception track	F14	ExPe	External perception		Ego

(continued)

Table 6.1 (continued)

AL2	Abstraction level 2 (AL2)	AL1	AL1	Abstraction level 1 (AL1)	Layer affiliation	Part of the psychic apparatus function
					Layer 3 (psychic apparatus)	
		F46	MTPe	Memory traces for perception	Layer 3 (psychic apparatus)	Ego
		F37	pRfP	Primal repression for perception	Layer 3 (psychic apparatus)	Id
		F35	EorC	Emersion of repressed content	Layer 3 (psychic apparatus)	Id
		F45	DoPI	Discharge of psychic intensity	Layer 3 (psychic apparatus)	Id
		F18	CoQP	Composition of quota of affects for perception	Layer 3 (psychic apparatus)	Id
DT	Defense track	F63	CoEV	Composition of emotion (to a vector)	Layer 3 (psychic apparatus)	Ego
		F55	Spa	Superego proactive	Layer 3 (psychic apparatus)	Superego
		F7	Sra	Superego reactive	Layer 3 (psychic apparatus)	Superego
		F6	DMfD	Defense mechanisms for drive-wishes	Layer 3 (psychic apparatus)	Ego
		F19	DMfP	Defense mechanisms for perception	Layer 3 (psychic apparatus)	Ego
		F71	CoeE	Composition of extended emotion	Layer 3 (psychic apparatus)	Ego
TT	Transformation track	F21	TtSP	Transformation to secondary process (perception)	Layer 3 (psychic apparatus)	Ego
		F20	CoFS	Composition of feeling (to a scalar)	Layer 3 (psychic apparatus)	Ego
		F8	TtSD	Transformation to secondary process (drive-wishes)	Layer 3 (psychic apparatus)	Ego

(continued)

Table 6.1 (continued)

AL2	Abstraction level 2 (AL2)	AL1	AL1	Abstraction level 1 (AL1)	Layer affiliation	Part of the psychic apparatus function
RT	Reasoning track	F69	EopG	Extraction of possible goals	Layer 3 (psychic apparatus)	Ego
		F23	foAt	Focus of attention	Layer 3 (psychic apparatus)	Ego
		F61	Loca	Localization	Layer 3 (psychic apparatus)	Ego
		F51	RCWf	Reality check wish fulfillment	Layer 3 (psychic apparatus)	Ego
AST	Action selection track	F26	DeMa	Decision making	Layer 3 (psychic apparatus)	Ego
		F52	GoiA	Generation of imaginary actions	Layer 3 (psychic apparatus)	Ego
		F53	RCAP	Reality check action planning	Layer 3 (psychic apparatus)	Ego
		F29	EoiA	Evaluation of imaginary actions	Layer 3 (psychic apparatus)	Ego
		F30	MoCo	Motion control	Layer 3 (psychic apparatus)	Ego
IT	Imagination track	F47	TtPP	Transformation to primary process	Layer 3 (psychic apparatus)	Ego
AT	Action track	F67	BoER	Bodily emotion-reaction	Layer 2 (neuronal symbolization layer)	
		F68	AcfG	Actuators for glands	Layer 1 (neuronal layer)	
		F31	NDAC	Neuro-desymbolization-action commands	Layer 2 (neuronal symbolization layer)	
		F32	AcfM	Actuators for muscles	Layer 1 (neuronal layer)	

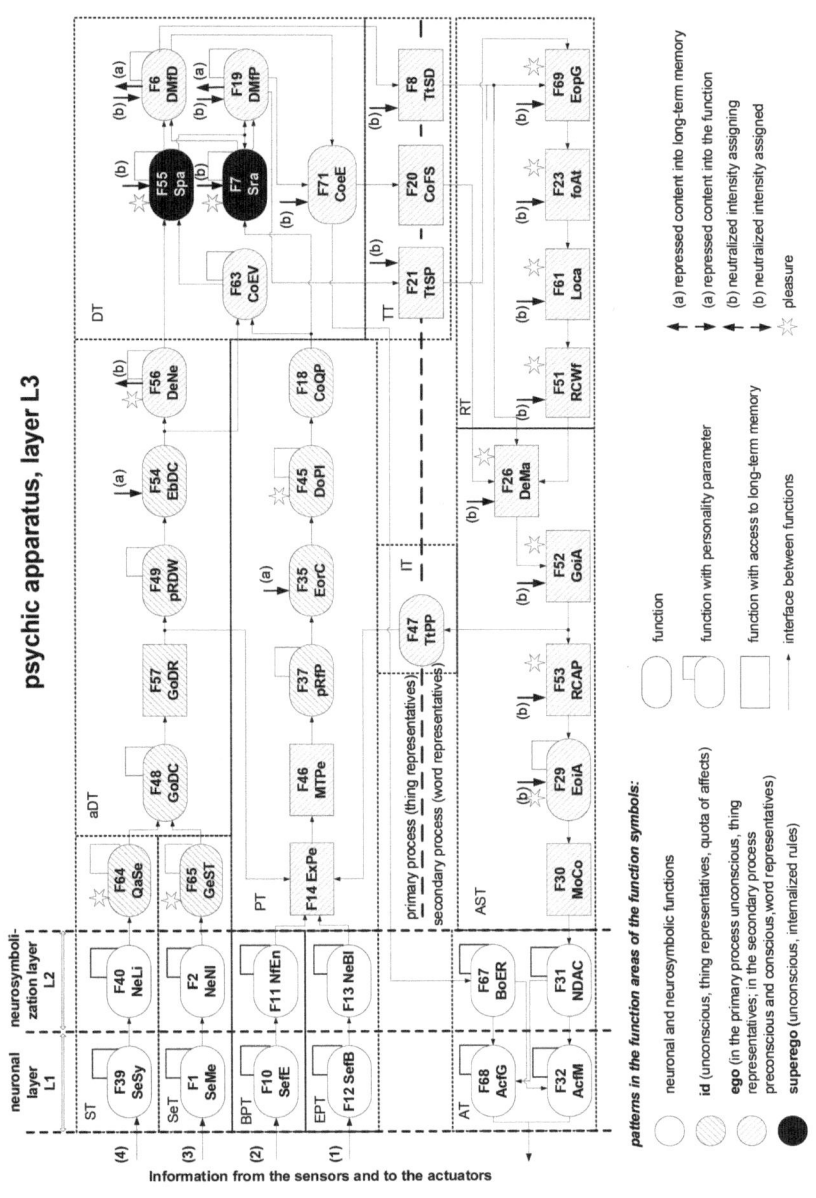

Fig. 6.14   The three-layer SiMA model in the lowest abstraction level AL1

functions, their possible effects on the various underlying memories, parameter settings, etc., which can be done independently of the other functions.[27]

The large dotted blocks in Fig. 6.14, labeled with the abbreviations ST to AT (compare the abbreviations of the functions in Table 6.1), which group together the respective functions Fxy, represent the *tracks* as defined in the previous chapter (abstraction level AL2). To simplify the graphical representation, the names of the functions in the tracks are given abbreviations from AcfG (actuators for glands, F68) to TtSP (transformation to secondary process (perception), F21) and additionally function numbers from F1 (SeMe) to F71 (CoeE).

The third type of defined functions are those that need to access personality parameters, for example functions F64 and F65. This raises a question that we have not yet been able to resolve in SiMA: What is the difference between the information in long-term memory and the information that represents personality parameters? This needs to be clarified from a psychoanalytic and perhaps also a neurological perspective. I find the problem difficult to grasp. It will have to do not only with memories, repressions, learning and some other processes, but perhaps also with inheritance through genetic developmental processes. This is where the relevant experts are needed. This does not yet play a role in the current simulation experiments in SiMA.

I have used the term *learning* here, which, like the term *intelligence*, I have strictly excluded from SiMA so far. To this date two further Ph.D. students have attempted to tackle this topic within SiMA. However, their work is not without its problems. From the point of view of SiMA, the terms of *learning* and *intelligence* need to be scientifically re-examined, i.e. they need to be harmonized axiomatically with all the other terms of the SiMA project. Why is this necessary? Consider Fig. 6.14. What functions are involved in processes and abilities that we call *learning* and *intelligence* respectively? According to the current state of research, we must assume that in principle all functions in the Ψ-organ are subject to developmental changes in the course of life, especially if we consider the human life cycle from the embryo to the elderly human or Homo sapiens in general in the course of its developmental history. For this reason, in the functional layer model, all functions, regardless of which layer they are assigned to (L1, L2, or L3), must be examined for *learning* and *intelligence*. How are learning and intelligence to be defined in each case? What are the goals and qualities of learning in each function?

If one takes a closer look at *learning*, the question arises: How should the abilities, qualities, or efficiency be measured for the psychic apparatus or its individual learning functions? First of all, reference values and benchmarks must be developed. For example, one might focus on the ability to learn to track individual objects better and better. Or learning to keep track of a group of people, or learning to

---

[27] When developing an abstracted model, the lowest level (= level AL1) is chosen so that the individual functions can be defined and described relatively easily and *independently of each* other on the basis of their *structure, parameters* and their *input* and *output variables*. This requires a simple definition of the interfaces between the functions.

solve mathematical problems, or learning to paint. If we go into more detail, each function must be analyzed in terms of its own learning process, and some of the reference values and benchmarks must be defined independently of the others.

And the term *intelligence* should be treated in exactly the same way. Intelligence must be defined in relation to each function, and when intelligence is applied to several functions simultaneously, or to the whole psychic apparatus, the appropriate reference values and benchmarks must be defined in each case. Whether it is learning foreign languages, dealing with mathematics, cooking, political activity, etc., each and every task requires different abilities and is therefore based on different principles of intelligence. Therefore, a principle of intelligence or learning in itself (without reference values and benchmarks) makes no sense.[28]

With the combination of all the basic functions of the SiMA model in Fig. 6.14, the classification of psychic intensity according to Fig. 6.8 may also become easier to understand. The valuation of the information, i.e. the assignment of different sizes (different valuation variables) of psychic intensity, must be carried out differently during the progression through the different functions. In this sense, brain researchers such as Antonio R. Damasio and Mark Solms emphasize the importance of making clear distinctions in order to properly formulate the process of *valuation of something*. Is the valuation based on a quota of affect, an emotion or even a feeling? Some are scalar quantities, the other vectors. In SiMA, this must be precisely formulated in the mathematical sense, as we are simulating the model, i.e. using natural scientific methods and laws as a basis. From layer L2, the neuronal symbolization layer, various amounts of quotas of affect are passed to the psychic apparatus (L3). These are scalars that are grouped into vectors. I think that such details will become clear in the further explanations of the individual functions.

It is important to first gain a better understanding of the individual functions in Fig. 6.14. F64 (QaSe: *generation of quota of affect for sexual drives*) and F65 (GeST: *generation of quota of affect for self-preservation drives*) are relatively simple functions. In them, the quotas of affect (scalar variables) of the *sexual drive representatives* (oral, anal, phallic) and the *self-preservation drive representatives* (the physiological sources are values of blood sugar, the muscular energy state, the rectal filling, etc.) are compiled from *drive tension representatives* of the neuronal symbolization layer. Which values of affect quota actually result from the input coming from L2 in the two functions F64 (QaSe) and F65 (GeST) will differ individually. This is why these two functions are assigned access to personality parameters (rectangular box on each oval in Fig. 6.14).

If in F64 (QaSe) a reduction in the quota of affect can be recognized (by comparing current with previous values) via incoming representatives (of signals from erogenously aroused zones), this represents a pleasure gain. In the subsequent function F63 (CoEV), where the emotion values are formed, this has a correspondingly positive effect, as well as in F20 (CoFS), where the feelings are formed. In Fig. 6.14, this function F64 is therefore marked with an asterisk.

---

[28] From this we can conclude: A categorical statement that one person is more intelligent than another is scientifically untenable.

The same applies to F65 (GeST). If the current representatives of homeostasis show better values than the previous ones, this is also recorded as a pleasure gain, which is represented by an asterisk in Fig. 6.14.

In F48 (GoDC: *generation of drive components*), the resulting representatives are split into more or less strong libidinal and aggressive representatives (also called drive components), depending on the personality. F57 (GoDR: *formation of drive representatives*) is shown as a rectangle, because this function searches for memories (also called memory traces) that promise a solution to satisfy the various drive tensions. This can be seen as a primitive mode of operation of the psychic apparatus, which operates purely on the pleasure principle and is tantamount to hallucinatory wish-fulfillment. It works according to the precept: Which *drive aim* and which *drive object* gave me the greatest pleasure in the past? These memories have been valuated in terms of the quota of affect in the past and are incorporated into the new valuation of the quota of affect. In this case, psychoanalysis speaks of the cathexis of the aim of drive and the object of drive. These current drive-wishes are passed to function F49 (pRDW: *primal repression of drive-wishes*) and are also received by the perception track via function F14 (ExPe: *external perception*). Function F49 (pRDW) is now required to perform a remarkable feat. Of course, the drive-wishes also associate with repressed drive-wishes, but these must not become conscious. This must lead to conflicts. The psychic apparatus has developed many possible solutions to this enormous problem. For example, the object of the drive is exchanged for another, or the valuation, i.e. a certain value of the affect quota, is shifted to another aim of drive, etc. The psychic apparatus is very, very flexible in such cases.

The following function F54 (EbDC: *emersion of blocked drive content*) is less complicated, since its contents are not primally repressed, i.e. they are not subject to a general prohibition on becoming conscious. Thus, if the incoming drive content corresponds in some way to a *repressed drive representative* present in long-term memory (arrow (a) pointing to the function), it is simply appended with the valuation of the repressed drive representative.[29] In this way, repressed content is able to pass through the defenses in a slightly altered context. The well-known phrase "forget that old bad story" is thus reduced to absurdity. Even if they are rarely (or never) consciously perceived, repressed drives continue to influence current actions. The repressed content is thus constantly activated and kept alive. It is not so easy to really forget, especially when you consciously set out to do so.

F56 (DeNe: *desexualization/neutralization*) is again a complex function. An individual *degree of neutralization* for each person determines how much of the *psychic intensity* (affect, basic emotions, extended emotions) is neutralized. This part, the so-called *neutralized psychic intensity*, is no longer available for direct drive satisfaction and is allocated to the valuations of other (higher cognitive) functions.[30]

---

[29] Repressed contents are stored in long-term memory, into which they have been displaced by the defenses (functions F6 (DMfD) and F19 (DMfP)).

[30] For psychologists and psychoanalysts, I would like to point this out again: If we disregard mystical explanations, the psyche is purely an information process. In the SiMA project, therefore, I have always firmly refused to allow the term psychic energy. In SiMA we call it *psychic intensity*. Because if you stick to the psychoanalytical concept, you could be misled into thinking that there is

These are ego functions, but not all of them. They are only those shown in Fig. 6.14 with an arrow pointing inward. They are mostly secondary process functions. Their *neutralized psychic intensity* is increased. In Fig. 6.14, therefore, the arrow (b) of F56 (DeNe) points outward.[31] This enhancement of psychic intensity allows for increased cognitive performance. It is the prerequisite for being able to go through complex thought processes. The ability to raise a lot of *neutralized psychic intensity* in F56 (DeNe)—that is, to greatly reduce the value of sexual intensity (i.e. that psychic intensity that enters via F64 and so the sexual drive track) in F56 in order to increase the neutral psychic intensity in other ego-functions—can thus also be seen as a measure of a person's adult development. However, if deep sexual needs are no longer satisfied, this can lead to personality disorders, or the person may appear factitious to others.

The *degree of neutralization* is one of the personality parameters that characterize an individual.

A second major pool that requires *neutralized psychic intensity* is the defense track (DT). It needs it to change or even repress contents (objects of drive or even aims of drive).

The question is: How does pleasure gain, i.e. *drive discharge*, occur? And how can it be modeled? SiMA is based on the following hypothesis: The psychic intensity distribution initiated by F56 (DeNe) in favor of ego functions in particular makes it possible to force the processes in such a way that they can lead to a tremendous increase in performance. In this way, the high pleasure of the sexual drive track is redirected to *sublimation*.

How is the *neutralized psychic intensity* distributed? This is to be imagined as follows: The individual functions that require *neutralized psychic intensity,* have connections to F56 (DeNe) to inform it of the urgency of the need for this *neutralized psychic intensity*. Depending on the personality of the person, F56 (DeNe) then raises the amount of valuation in the various functions more or less, especially depending on the requirements of the functions themselves. However, *it is assumed that a certain absolute value of the psychic intensity to be distributed is never exceeded.* SiMA simulations will have to provide evidence for this hypothesis.

Let us switch to the perception track (PT) in Fig. 6.14. It receives its information from the *body perception track* and the *environment perception track* of the neuronal symbolization layer. Identified patterns such as objects, movements, smells, phonemes, etc. are fed into function F14 (ExPe: *external perception*). There they are

---

an "energy flow" here. I repeat myself: nothing "flows" in the psychic apparatus. Valuations such as psychic intensity can be *assigned*. But they are not subject to physical laws. Information can only be processed on the basis of Eq. (5.1) (Sect. 5.1). However, the laws of information engineering must be strictly decoupled from the laws of physics. This is a fundamental definition in the theory of information engineering.

[31] The arrow must not be misunderstood. Valuations such as quota of affect, emotion or the umbrella term psychic intensity are purely informational variables. They are not physical quantities, so they cannot flow. An arrow pointing outwards from a function only means that this function influences the valuation of other functions, an arrow pointing inwards to a function means that the valuation of this function is influenced.

compared with internally stored patterns, which can also be fantasy patterns (in Fig. 6.14 the arrow comes from the secondary process below, specifically from F47 (TtPP: *transformation to primary process*)). The output variables of F14 (ExPe) receive a valuation by quotas of affect via the remembered patterns as well as via the current situation of the drive track.

F46 (MTPe: *memory traces for perception*) attempts to connect adjacent patterns identified in the external world in order to initiate new associations. As with all L3 functions, the information is stored in short-term memory. This is relevant because as a consequence the agent does not immediately forget the pattern if, for example, an object is briefly lost from view. I mention this detail because it is a good example of how modeling errors can quickly become apparent in SiMA. It happened to us that we initially forgot to store the value of this function in short-term memory. In the simulation, the agent then showed strange behavior that we know this from infants: The agent would see an object, and if something then visually came between him and that object, such as a tree, the object was no longer there for him—he immediately "forgot" it.

F37 (pRfP: *primal repression for perception*) and F35 (EorC: *emersion of repressed content*) are functions that can be considered equivalent to F49 (pRDW) and F54 (EbDC) of the applied drive track, except that in this case the information represents patterns of external perception.

Pleasure can be generated via the function F45 (DoPI: discharge of psychic intensity) if the satisfaction of an achieved aim of drive is attained, but this is not achieved via a physical improvement (pleasure gain, e.g. via a more balanced homeostasis), but via an event experienced in imagination. Such an event could be the sharing of a piece of meat, if that was the agent's goal and he was successful. In such a case, a pleasure gain occurs (see asterisk of F45 (DoPI)). A personality parameter is also assigned to this function.

In F18 (CoQP: composition of quotas of affect for perception), quotas of affect of the various patterns (Fig. 6.15) present are combined. In this way, the current quotas of affect for perception are mixed with repressed and primal repressed quotas of affect as well as other associated memory-traces, which makes the past more or less alive again and again with regard to the quotas of affect. It is therefore not so easy to forget this content, as I have already mentioned. In F18 (CoQP) there is yet another synthesis. Thing representatives from the current perception as well as memories activated with them are compared with equivalent connections of the drive representatives. As a result those connections are treated as equivalent, which are based on the same thing representatives and in which the three drive components (drive aim, drive source, and drive object) match.

The defense track (DT) that follows in the progression of information in Fig. 6.14 represents an extremely complex function. In SiMA, we have analyzed and synthesized it for years, and yet only an extremely limited part of the insights developed in psychoanalysis to this date could be taken into account in the dissertation (Gelbard 2015) and in the publication (Dietrich et al. 2015, p. 128). In contrast, I consider this function extremely important for AI. In order to achieve similar abilities and qualities in machines as in humans, the mechanisms and principles of *defense* of

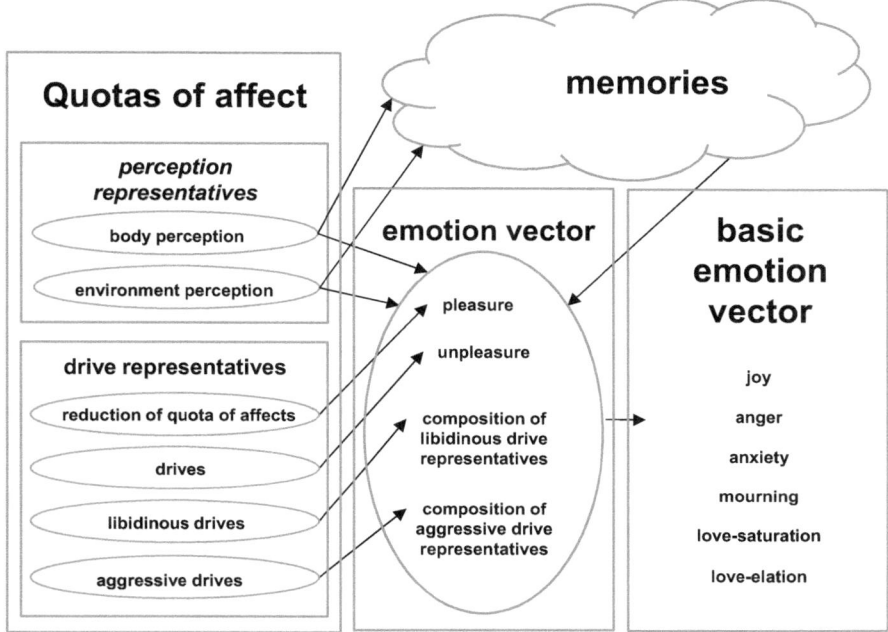

Fig. 6.15 Development of basic emotions [adapted from Dietrich et al. (2015, p. 129)]

the psychic apparatus must be studied intensively.[32] Conflicts are a precondition for the activation of defense. They are almost always present. Imagine the demands that homeostasis or the sexual drive are constantly making, but that a person's current social situation and superego do not allow at the given time. I often mention the following situation as an example of an external demand of reality. I am giving a lecture that I am enjoying immensely. At the same time, my stomach is rumbling. I see pleasant people in front of me with whom I would like to interact. This means that my psychic apparatus is confronted with many contradictory wishes at the same time. In this sense, I think there is hardly any situation that allows for a conflict-free psychic apparatus. People have to be able to deal with conflicts all the time. And for that they need their defenses.

---

[32] I have mentioned this thought before, but because of its enormous importance, I would like to emphasize it again here. To put the advantage of the primary/secondary process organization in a nutshell: The primary process allows, like a massively enhanced brainstorming, to conjure up as many possible solutions as fantasy allows (based on the principle of gaining pleasure and avoiding unpleasure), while the secondary process has the task of selecting and refining those actions that offer the greatest real chance of achieving the set aim. However, the secondary process can only do this if it is not overwhelmed with too many fantastic solutions and thus becomes incapable of acting. To achieve this goal, defense plays a central role. Shock experiences are a good example of this. At the right time, the defense must reject all information that could incapacitate a person at a critical moment. Often the only remedy is complete repression.

Two key functions are crucial for inner conflicts: the *superego proactive* (F55 (Spa)) and the *superego reactive* (F7 (Sra)). While the reactive superego requires only the emotions of the perception track as input, the proactive superego processes a compilation of all emotions, which requires another function of the defense track: *the composition of emotion* F63 (CoEV). This line of thought leads back to Fig. 6.8, the classification of psychic intensity, i.e. all the evaluation variables of the psychic apparatus. This is an extremely difficult topic, since psychoanalysis distinguishes between quota of affect and emotion, but does not further differentiate between emotions and feelings. Therefore, as already mentioned, SiMA relies for this question mainly on Antonio R. Damasio's definitions, since the individual functions in Fig. 6.14 require more precise and differentiated definitions, up to the concepts of extended emotions. In order to explain why, I would have to delve deeper into the subject and thus into mathematical descriptions, which would defeat the purpose of this book and probably bore the general reader. Interested readers are referred to the various and especially the most recent dissertations, for example the work of Schaat (2016), which deals with these relations in considerable detail. There it is also explained that not all valuation variables can be simple scalars, i.e. variables with a numerical value, but must be partially described by vectors that are composed of scalars. The quotas of affect, which are scalars, are the origin of all valuation variables. The different emotion vectors (emotions, basic emotions, and extended emotions) then develop from them. For the secondary process, the scalar of feeling is determined from the quotas of affect and the various emotions.

I will try to give you a rough overview of the development of the basic emotions using Figs. 6.8 and 6.15. The individual emotion variables such as pleasure, unpleasure, the sum of all libidinal and the sum of all aggressive drive representatives are derived from the quotas of affect of the drive representatives. For the resulting basic emotions, the valuation variables of the quotas of affect of the perception representatives from the perception track are added (note that the quotas of affect of the current perception are mixed with associated valuations from memory in the perception track).

Various basic emotions are mentioned in the literature. In SiMA we have primarily drawn from Mark Solms (2013) and Jaak Panksepp (1998, p. 51). It is often claimed that they are innate and not learned. However, the scientific evidence does not seem to be very strong. We in the SiMA project do not want to get involved in this discussion, but we tend to believe—if we take the model in Fig. 6.14 as a basis—that the psychic apparatus has to "work them out" what in our common parlance is equivalent to "learning". This train of thought was one reason for us to place the formation of basic emotions only in F63 (*the composition of emotion* (CoEV)), i.e. relatively far back in the primary process. For the time being, this is not crucial for modeling the psychic apparatus,[33] but it is very important which basic emotions we have defined in SiMA. These are, according to drive-wishes (Dietrich

---

[33] I am aware that there may also be fundamental errors in the present model, but such errors will become apparent in future experiments, which is why it is also necessary to simulate the SiMA model in different ways.

et al. 2015, p. 129), joy, anger, fear, grief, happiness satiation, and happiness elation, defined according to the rules of a dominance of:

- pleasure causes joy,
- unpleasure causes fear,
- unpleasure with a high libidinal ratio causes grief,
- unpleasure with a high aggressive ratio causes anger,
- pleasure with a high aggressive ratio causes happiness elation and
- pleasure with a high libidinal ratio causes satisfaction.

I think that these definitions clearly show the fundamental difficulties of such a classification. Since it cannot be measured objectively in a person, but must be based on the subjective investigations (experiments) of psychoanalysts, neurologists, and behavioral scientists. We must adopt those of their assumptions that seem most convincing to us in relation to our model in SiMA. Their correctness and robustness must then be verified in the simulation experiments. This means that some modifications and redefinitions will certainly have to be made in the future. However, this is a normal procedure in natural sciences and especially in engineering. Only extensive experiments will allow a certain degree of certainty in the definitions of concepts. It is not without reason that international standardization commissions allow the definitions of terms to be modified every 4 years.

Back to Fig. 6.14 and the superego functions. Function F7 (Sra: *superego reactive*) is the function that is easier to describe. It compares social superego rules with the content of the applied drive track. This generates drive tension, i.e. unpleasure, which can be best accepted when it is supplied with as much neutralized intensity as possible. The resulting information is fed to defense functions F6 (DMfD: *defense mechanisms for drive-wishes*) and F19 (DMfP: *defense mechanisms for perception*).

The function F55 (Spa: *superego proactive*) is not as simple as F7 (Sra). It does not need to be stimulated by drive representatives. Depending on the personality parameters, the neutralized psychic intensity assigned to it determines when it becomes active in which situation. In certain situations, it calls forth demands such as: "Do a good deed every day!" (Dietrich et al. 2015, p. 132). If there is a conflict with the information in the perception track, it is the task of the defense functions F6 (DMfD) and F19 (DMfP) to solve this problem.

The major part of the task of the defense track (DT) thus falls to the two functions F6 (DMfD) and F19 (DMfP). On the one hand, drive-wishes are to be satisfied, on the other hand, superego rules must not be disregarded. As mentioned in Sect. 3.3, this is often not directly solvable, not even through compromises, which is why the psychic apparatus has developed mechanisms that are not so easy to grasp. And I make the claim: The more imaginative a person is, the more differentiated defense mechanisms they have at their disposal. Three principles can be distinguished (Deutsch 2011, p. 95):

Drive-wishes can

(a) pass through the defense functions unchanged,
(b) pass through them partially or completely modified, or
(c) they can be partially or completely repressed by them.

The result is expressed in the valuation of the fed forward information. The task of function F71 (CoeE: *composition of extended emotion*) is then to extract from it the extended emotions (see the corresponding table in Sect. 6.3.2), which represent an accentuation or coloration of the emotion vector. They are (in addition to the basic emotions) quantities from which the emotions are formed in the secondary process.

Note that an output of function F71 (CoeE) also goes directly to layer L2, so this progression of information cannot be directly influenced by the secondary process, i.e. it cannot be consciously influenced. This expresses itself in phenomena that everyone knows. Pity, shame, pride, fear, if they are strong enough, are expressed directly in a person's facial expressions without the person being able to consciously suppress them when they arise—unless you are a trained actor.[34] The information is channeled directly through L2 the layer of neuronal symbolization to the muscles and glands.

The transition from the *primary* to the *secondary process* is an enormous step for the psychic apparatus. It occurs exclusively through defense. The SiMA model provides three functions of the primary-secondary transformation track (TT). Word representatives are added to the thing representatives of the primary process, an essential component for psychic contents to become preconscious or even conscious. The function F20 (CoFS: *composition of feeling* (to a scalar)) receives a list of extended emotions with the sizes of the current ratings from F71 (CoeE). F71 (CoeE) requests psychic intensity from F56 (DeNe), the size (see for this Fig. 6.18) of which is determined by personality parameters. The assignment variables determine which current values the extended emotions will ultimately receive, from which the value of the feeling is then determined. The interaction of the basic emotions, the emotion vector, the extended emotions and the feeling provides the weightings for the possibilities of how the various drive-wishes that have passed the defense could be satisfied. Function F26 (DeMa: *decision making*) selects a small subset of these possibilities to be forwarded.

But let me return to the *transformation track* (TT), i.e. the transformation from the primary to the secondary process. So far, only patterns (visual, static images, movements of objects, smells, sounds, etc.) have been discussed, which are processed in the primary process. They are interwoven as thing representatives into networks that can be associated in the primary process. The secondary process is characterized firstly by the fact that word representatives are added to these thing representatives, secondly by the fact that the reality of the external world plays an important role, and thirdly by the fact that causality is taken as a basis and thus the temporal sequence of a process is a decisive factor. What happened in the past can only emerge as a presentation of how a scene might have taken place. A person has

---

[34] As a computer scientist, this leads me to an interesting question that I derive from Fig. 6.14: Can actors immerse themselves in another world in such a way that their primary process lives in it?

to accept the now and the present moment. In the secondary process, past events are memories and thus the past (unlike in the primary process that does not consider time), and this enables me to imagine events in the future to the same extent. In other words, the secondary process allows a person to plan and think logically. A person goes from place A to place B, then to place C, and so on. In SiMA we call this kind of imaginary sequence an *act*. Acts are process sequences, i.e. also longer sequences of movements, which the psychic apparatus develops in an abstract way on the basis of patterns. For example, if I decide to get a coffee, I have to develop a *logical* plan of the activities I have to carry out before I can finally hold the coffee in my hand. In terms of the previous Sect. 3.2 the secondary process—by which I mean the individual functions in the secondary process—has the ability to develop logical process sequences on the basis of patterns (symbols).

The task of function F21 (TtSP: *transformation to secondary process (perception)*) is not as simple as it might seem at first glance. In the primary process, all thing representatives are linked in various lists, which are associated (i.e. activated) when, for example, a corresponding object is fed in as an information variable via a perception. The thing representatives that passes through defense are fed into *word representative lists* whose words are linked to similar thing representatives. The words found are again linked to networks, but these should be kept as small as possible. This is where the neutralized psychic intensity comes into play, which must be requested from F56 (DeNe). If a low psychic intensity is assigned (arrow (b) pointing in Fig. 6.14), the number of associations will be kept low. If a high amount is assigned to F21 (TtSP), the number of associations can be quite extensive. As is often the case in nature, small changes (in this case in psychic intensity) have here very large effects.

Function F8 (TtSD: *transformation to secondary process (drive-wishes)*) receives drive-wishes permitted by the defense. These are drive representatives with corresponding associations. From this information, drive aims are associated in F8 (TtSD) and added to the information along with word representatives. The extent of the associative capacity in turn depends on the psychic intensity provided by F56 (DeNe).

This brings us to the reasoning track (RT). The information from the inputs of the first function F69 (EopG: *extraction of possible goals*) is used to compare the perceived solution possibilities with the drive-wishes and drive aims within the *drive* and *defense tracks* and to filter out what is not possible, i.e. what violates the premises of the secondary process (causality, temporal sequence, etc.). If there is no drive aim that can be considered, further drive aims are associated given that a sufficiently high level of psychic intensity is available.

The number of options for satisfying different needs is enormous. An optimal decision can only be made if the flood of information is once again strongly filtered, which is done by function F23 (foAt: *focus of attention*). Selected patterns are overcathected[35] via the neutralized intensity, which is a prerequisite for information

---

[35] Overcathecting means that selected patterns are valued more strongly, i.e. in the case of function F23 (foAt), they are given an additional emotional value.

to become conscious. More generally, the psychic apparatus directs its focus to certain solutions and thus suppresses less important ones. If I am in the jungle and have to expect a tiger, I concentrate fully on its appearance. Butterflies and flowers, or perhaps bicycles and washing machines, which may be associated, are not the current issue.

It is important to locate yourself in two ways. First, you need to know where you are, and second, you need to know how to get from point A to point B. To do this, you need to develop an internal map. For this purpose, certain important locations are stored as patterns. In the SiMA model, the function F61 (Loca: *localization*) has been defined for this task.

Localization is not only a complex process, it also has to be seen in a larger context. When I left my company to go to university, my boss at Telenorma gave me something special as a farewell present, which he himself found funny: a workshop in Switzerland to combat forgetfulness. At first, I was not really able to get too excited about it, but then I became fascinated by it. The speaker used to be a stenographer in the German parliament. He explained to us that he was going to teach us the method of memorizing a text like the Bible, as it used to be customary, so that we would not forget it again. Of course, being a very forgetful person, I was very skeptical and doubted his statement. The lecturer referred to Gaius Julius Caesar who taught this method 2000 years ago. At the beginning of the 3-day workshop, we had to make a shopping list. Usually, after I completed such a list, I had only fragments of it in front of my mind's eye. But back then it was unexpectedly completely different. After using the method, we threw the notes away, but for many years afterwards I was able to list all the items on the shopping list in order. It was only my lack of engagement with the list that caused it to fade more and more from my memory.

How does this memory method work? It starts with these instructions: Think of a route to a place that has been "burned" into your memory. For me: I sit in my bed. From there, I look for another important place, like the nightstand. From there I look around and find the next visible important place, the chair by the door. In this way, I establish a fixed route of places that are personally important to me. Every fifth and tenth place has to be a standout. My shopping list started with lemons. I had to mentally squeeze them in bed, smell them and feel the sticky moisture. Next was the milk for my coffee. I had to mentally pour it over the nightstand, smell it, touch it, maybe hear it dripping, and so on. If you start with a shorter list, say five or six objects, and you check the list after 3 min, then after 15 min, then after an hour, and so on, you cannot forget the individual objects. I still remember the first objects after about 40 years. The interesting thing is that you do not forget anything along the way. If you follow the path once you have laid it down, you remember every single point. And if you practice intensively, you can even jump to the outstanding points of the fifth and tenth position. What is the principle behind this? The principle of association lists and the networking of all the lists of different patterns that are stored in the psychic apparatus, patterns of the sense of sight, the sense of smell, the sense of touch, and so on.

What does this story tell us? Several things that are crucial to SiMA. First, it confirms the theory that information is stored and retrieved in the Ψ-organ via associations, i.e. networks of lists. Second, information that involves an interplay of different sensory perceptions guarantees a more robust retrieval by linking the different memory lists. Third, it also shows that spatial (three-dimensional) thinking plays a central role in the psychic apparatus and cannot simply be seen as independent of other activities. Fourth, familiar places seem to be "burned" into the Ψ-organ, although we know that the neural system also renews itself again and again in the course of life. And the principle of preserving information despite the death of synapses certainly applies to the same extent to other activities of the psychic apparatus, not only to localization: If strong emotions play a role in storing information, it is no longer so easy to "forget" information. By consciously wanting to forget, one achieves the opposite, since the synapses, on which the memory relies, are activated.

In summary, it should be noted: Spatial presentations are a central principle of the psychic apparatus, especially for unconscious processes. Thus, the relationships of localization, as developed in Brandstätter (2020), can only represent the beginning of such a consideration.

The ability to maintain a local all-around awareness must be learned as a baby. Function F61 (Loca: *localization*) is responsible for this. Everyone knows the game with a baby where you "hide" behind a blanket and delight the baby when you suddenly reappear.

For this purpose, not only the different patterns of locations are stored in long-term memory, but also entire sequences of actions to get from A to B—as I said, in SiMA we call them acts.

Two aspects still need to be considered with regard to F61 (Loca): Firstly, localization certainly does not work with algorithms like today's electronic positioning systems, because we do not have an arithmetic unit in our Ψ-organ. The basic principle is different. Remember the ball catching robot (Sect. 3.2). The psychic apparatus memorizes places with a valuation,[36] they are called *landmarks*. And if I want to get from A to B, I remember (valuated) places in between, which is why based on my valuation I ultimately end up on the path I want to take. It is logical that the primary process often plays a crucial role here, because associations are awakened for all the different places that come into the "inner" eye, the valuations of which are included in the selection to be made.

The second aspect concerns the naming of functions in the psychic apparatus. In SiMA we call F61 (Loca) the function of *localization*. It must be clear that the whole process of localization of the psychic apparatus cannot take place only in this one function F61. Localization involves many functions, but in F61 the decisive process of localization takes place. And this principle applies to a similar extent to almost all other activities of the psychic apparatus. All activities such as speaking, thinking,

---

[36] The ball-catching robot stores and remembers valuated trajectories (Pongratz 2016) and does not calculate the ball's path.

ɔlaying music, boxing, deciding, concentrating, etc., require the activity of many ẕunctions of Fig. 6.14, but some functions take over a central moment of the task in each case.

Before a preliminary decision can be made as to which two or three actions are possible in principle, a reality check must first be performed. This task is performed by function F51 (RCWf: *reality check wish fulfillment*). It checks whether the selected solution options can actually be realized with the perceived objects. Does the external reality match my imagined process or is it pure fantasy?

The preliminary decision as to which drive-wishes can be satisfied is made in function F26 (DeMa: *decision making*), i.e. depending on which decision promises the greatest level of pleasure and lowest level of unpleasure. This is why the input variables are derived from the aforementioned function F51 (RCWf), but also from function F8 (TtSD) and above all from function F20 (CoFS), in which the extended emotions were developed. The possible drive aim that come into question are not determined. They are passed on as a list.

What was previously hardly imaginable has now been recognized and substantiated by science: Before humans actually perform an action with their bodies, they always (unconsciously) carry out various trial actions in their thoughts (we technicians speak of simulations in such a case), before deciding on the one that promises the greatest pleasure gain and the least unpleasure. Mark Solms says: "Thinking is acting without acting" (Solms and Turnbull 2002, p. 281). This is the task of function F52 (GoiA: *generation of imaginary actions*), which of course takes place in an "imperceptible" amount of time and, above all, unconsciously.

The resulting information is not only passed on to function F53 (RCAP: *reality check action planning*) in order to arrive at a bodily action execution, but is also fed back to the primary process via function F47 (TtPP: *transformation to primary process*), which above all influences function F46 (MTPe: *memory traces for perception*) in the perception track.

This feedback is very important in several ways. In principle, feedbacks have enormous advantages. We would not be able to grasp our cup of coffee properly if ɾot via our eyes and our sense of touch information was fed back in order to make fine adjustments. It is only through this feedback that a precise grip is possible. Scientists in robotics know the problem all too well. However, any feedback has the ɔisadvantage that it can lead to instability in the entire process: a central issue in control and automation engineering. Feedback can lead to behavior that is difficult to ɩnderstand (Dietrich et al. 2015, p. 30), which is why SiMA deliberately avoided the ɱany feedback options in the initial development phase and only considered this one via function F47 (TtPP).[37] It could not be avoided, since the model would have ɔtherwise massively violated psychoanalytic theory. There is no question that in the fɩture (slowly, step by step) further feedbacks will have to be integrated in order to be able to better and better approximate a person's behavior using simulation engineering and thus understand it increasingly better.

---

[37] In the next chapter I will return to the topic of feedback, since it plays a crucial role there.

Through the transition from the secondary to the primary process (in the transition from F47 (TtPP) to F14 (ExPe)), the word assignments are again separated from the information; what remains are valuated thing representatives. The data structures of the thing representatives in turn lose the ability of becoming conscious.

In F53 (RCAP: *reality check action planning*) of the action selection track, the actions played through in thought are checked against the associated factual knowledge, i.e. also with regard to causal relationships. The final decision is made in F29 (EoiA: *evaluation of imaginary actions*), again according to the principle: Which of my planned actions brings the most pleasure, which the most unpleasure?

The result is fed to F30 (MoCo: *motion control*). It must transform the next step of action into representatives of bodily movement.

Since SiMA focuses only on layer L3 (psychic apparatus) and layers L1 (hardware/neuronal system) and L2 (neuronal symbolization) are more or less represented by dummies—i.e. the agent has a very limited body in comparison to a human—the task of the functions of layers L1 and L2 in the simulation program can still be regarded as extremely rudimentary. When the simulation package will later be integrated into a robot, the task of this function will become significantly more important. I see its realization as a challenge for us engineers, on which today's humanoid robot technicians are already working diligently. I see open scientific questions regarding the two layers L1 and L2, especially regarding perception, which will be addressed briefly in the following chapter. The last four functions will therefore be dealt with in a concise manner. The enormous complexity behind the functions of layer L2 is shown by Hawkins (2004), but also in the dissertation of Velik (2008). This also means that behind the functions of layer L2 and layer L1, shown in Fig. 6.14, one has to imagine numerous sub-functions, similar to those in layer L3.

The *emotional state* (which is not the same as the *state of feeling*) is reflected in bodily reactions through F67 (BoER: *bodily emotion-reaction*). This is the only function in this track that does not get its information from the secondary process, but via F71 (CoeE) directly from the primary process. It desymbolizes the received information and activates glands via F68 (AcfG: *actuators for glands*) and muscles via F32 (AcfM: *actuators for muscles*). One can experience its functioning when, for example, the heart rate increases in stressful situations or when we blush when we are angry. All these reactions cannot be consciously inhibited because F67 (BoER) is directly activated by the primary process via F71 (CoeE). They are individualized, which is why personality parameters are assigned to these four functions of L2 and L1. F31 (NDAC: *neuro-desymbolization-action commands*) desymbolizes all action commands to activate the various muscles and glands of the body. F68 (AcfG) and F32 (AcfM) represent the neurological (hardware) functions that activate the muscles and glands.

6.3.6   L2 and L1

Although layers L2 and L1 are not discussed in detail, I would like to give a rough idea of how they can be described in the overall model of the Ψ-organ. Jeff Hawkins was the first to propose principles of a solution that made sense to me and answered crucial questions (Hawkins 2004, pp. 117–118). However, he does not address layer 3 (the psychic apparatus). Rosemarie Velik, by contrast, was the first to take up the topic of layers L1 and L2 in a comprehensive way in connection with all three layers (Velik 2008). Her publications and especially her dissertation show how colossally complex the simulation of layer L2 is, and that it can only be carried out concretely if extensive results of the simulation of layer L3 (the psychic apparatus) are already available. Only then can the interface i3 in Fig. 6.3 be sufficiently specified. Its specification is the prerequisite for the development of layer L2. However, Rosemarie Velik did not address the problem of *invariant representation*, a topic that is still of great concern to robotics and AI. Jeff Hawkins dealt with it in detail in Hawkins (2004), but unfortunately did not evaluate his hypotheses in the context of a simulation or a concrete implementation.

Here is a brief summary of the problem and the two solutions proposed by Rosemarie Velik and Jeff Hawkins.

Let us consider Figs. 6.3 and 6.4. The best way to explain the relationships is to describe the progression of information from layer L1 via interface i2 through layer L2 to interface i3. This was presented in (Dietrich et al. 2009, p. 105), and the first foundations for SiMA were laid in (Pratl 2006, p. 33). For example, the human optical system does not capture information like a camera. Even in the retina, different types of neurons filter different information from the visual input. Contrast is enhanced, shapes are recognized, as are objects that move towards you, etc. (Universitätsklinikum Freiburg 2019; NIH 2001). These are processes that can be explained and described in terms of purely physical and chemical processes, i.e. with the help of neuronal descriptions. All the information from the sensors and the resulting information from the lowest neuronal layers reaches the Ψ-organ and, according to Fig. 6.16a, forms patterns across many sub-layers of functional units, which can ultimately be described symbolically in layer L2 and can be imagined as a hierarchical list (Fig. 6.17). This means, for example, that the eye function in layer 1 (hardware/neuronal system) has the task of converting the light quanta into electrical information thereby sharpening the contrast, perceiving objects flying towards you, etc., and passing this information on to layer L2 at the interface i2. This layer interprets the various electrical information as patterns, which are mathematically described as symbols. These patterns associate previously learned, stored patterns and generate new patterns by merging them. Of course, this type of pattern formation applies in a modified form to all types of sensors, such as acoustic, olfactory, tactile, and so on. It is equally obvious that all information received from different sensors is ultimately combined, which is also implied by Fig. 6.17.

It goes without saying that a similar process also runs in the opposite direction, i.e. from the psychic apparatus via layer L2 to the neural system to control muscles and glands (Fig. 6.16b). So far, the explanations in Velik (2008) and Hawkins (2004) concur. However, the process is more complex than either scientist perhaps realizes. Figure 6.18 illustrates this in a very abstract way.

The human psychic apparatus (layer L3) would not be able to perceive all the information received through the various sensory organs. The details would be overwhelming. For this reason, layers L2 and L1 recognize patterns of characteristic

Fig. 6.16 (**a**) Information flow from the sensors to the psychic apparatus, (**b**) information flow from the psychic apparatus to the actuators

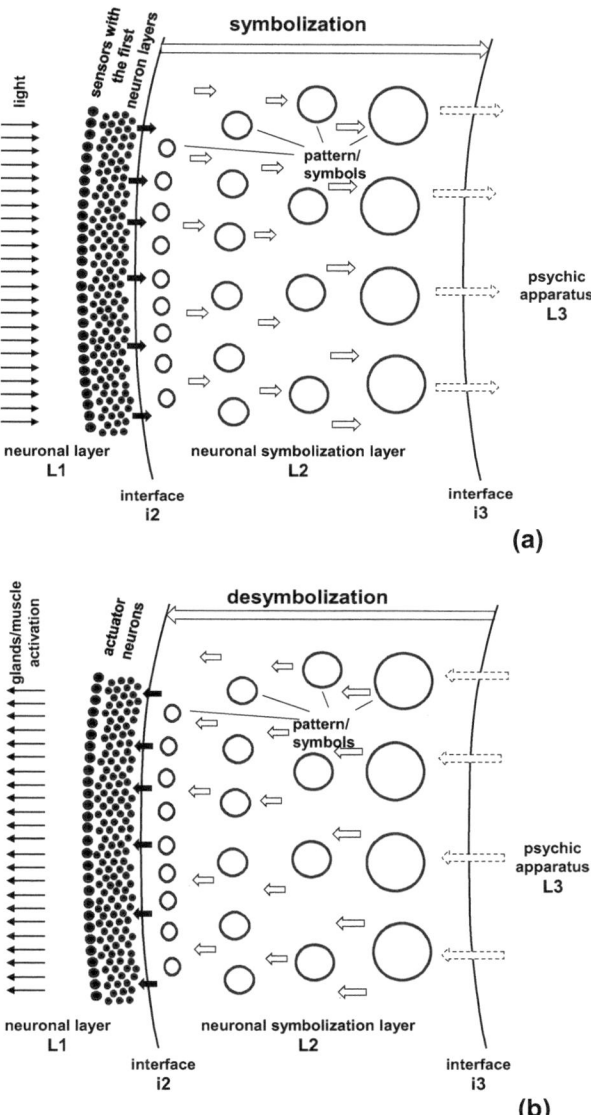

objects and relationships. In Fig. 6.18, these are, for example, the woman's eyes, mouth, or peculiarly shaped hair. From this, similar memories are associated with images from the past. This information is used to calculate an image,[38] the inner image that you believe is the object you see.

---

[38] Perhaps the reader can now understand why one is firmly convinced that one has seen certain things, even though others have contradictory memories. If one also considers that the human organs, including the neurons, are renewed every 10–15 years, it becomes evident that the

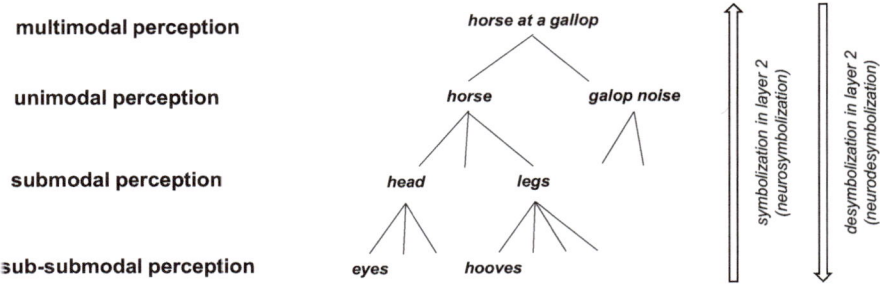

Fig. 6.17 Hierarchy of networked symbols/patterns

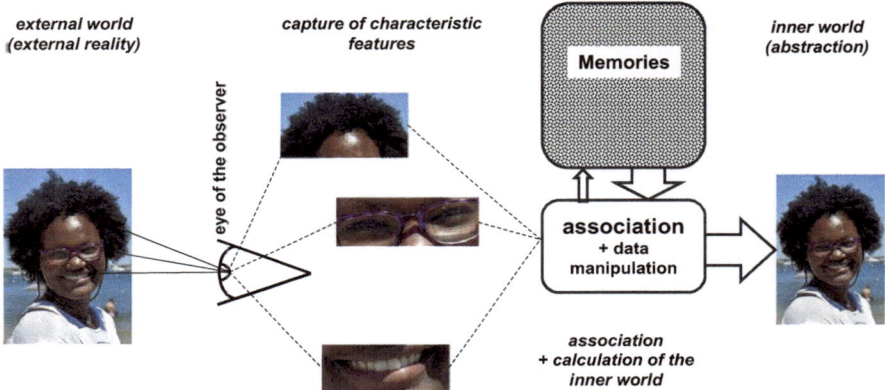

Fig. 6.18 The external and the inner world

Movements of objects in the millisecond range, such as falling balls or a scratching of your own skin, also generate patterns of information. A good example is certainly the characteristic gait of my wife. It is "burned" into my brain as a fixed pattern (symbol), just like my own characteristic movement when I bring the coffee cup to my lips. I cannot see this movement, but the movement of my muscles is stored in me in the form of patterns.[39]

---

information stored via synapses is also broken down. Consequently, the old information can only be retained by restoring it. This implies that current emotional valuations are ever mixed with new and old patterns, which may explain the unreliability of one's own memories over a longer period of time.

[39] I have had a tremor in my right hand since my teenage years. Several years ago, I took a β-blocker to stop the tremor, but not gradually as prescribed. I noticed an unusual phenomenon: The tremor was gone, but the hand no longer felt like it was part of my body. It felt as if it were an alien entity, as if it did not belong to my body. I realized that the patterns of my remembered hand movements did not correspond to the hand movements I felt. And the acquisition of these new patterns is a lengthy, a very, very lengthy process. This experience was particularly drastic for me.

The principle of perception therefore applies not only to the visual system, but also to all the other human senses. When you listen to music, what you consciously hear is a mixture of what you actually perceive through your ears and the sounds you have stored in your memory, something that Diana Deutsch has done a lot of research on, which has led to surprising and unpredictable results (Deutsch 2013; Klein 2015).

This way of understanding perception via layers L1 and L2, and conversely the activation of glands and muscles, might lead one to conclude that the process is, in principle, trivial. Unfortunately, this is not the case. The process of symbolization—and equally the process of desymbolization—operates via an extremely large number of feedback loops.[40] On top of that there are the ongoing associations of what has been learned. The number and diversity of networks of different patterns is enormous, especially when one considers the spatial and temporal components as well as the information from different sensors (Hawkins 2004, pp. 112–113, 115, 117–118). As mentioned above, Rosemarie Velik (2008) has found an answer to many of these questions. However, another issue, that of invariant representation, remains scientifically unresolved.

Let us return to the optical system to illustrate this. It is well known that the eye executes saccades at a rate of 3–5 per second (Hawkins 2004, pp. 94–95; Nature Reviews Neuroscience). This means that the retina in the eye is constantly sending new images at this rate. Consciously, however, we "see" a still, static image. How does the neuronal optical system combine the constantly changing images into a calm picture? And there is another phenomenon that has not been taken into account in Rosemarie Velik's explanations: Today's passport photos must adhere exactly to a prescribed head position, exposure, etc., otherwise they are not clearly legible. Even more impressive: when we see our friend, we can recognize him from all sides, but a computer program is hardly able to do this. Many images would have to be taken and stored in order for computer program to be able to recognize someone from different directions. However, it was previously unlikely that the computer would still be able to recognize it after many years of ageing, as a human often does. However, new AI systems are now more advanced in this respect.

This means that the human eye creates an *invariant representative* from the image of the retina, which certainly cannot be a pixel format, because otherwise the human optical system would have the same problems as robotics and AI, and it would have difficulties recognizing people from different viewing directions or even after many years. I find Jeff Hawkins's explanation (Hawkins 2004, pp. 77, 123) of how the eye and the nervous system behind it form an invariant representative brilliant. He assumes that optical perception is the responsibility of the smallest sensor surfaces (Fig. 6.19). The lowest functional unit of the optical system—a small set of sensors and the first layers of neurons behind them—is responsible, for example, for recognizing within a certain area edges, circular shapes, colors, objects approaching

---

[40]This is well documented by Rosemarie Velik's extensive research. Particular attention should be drawn to her detailed explanations in Velik (2008, p. 64).

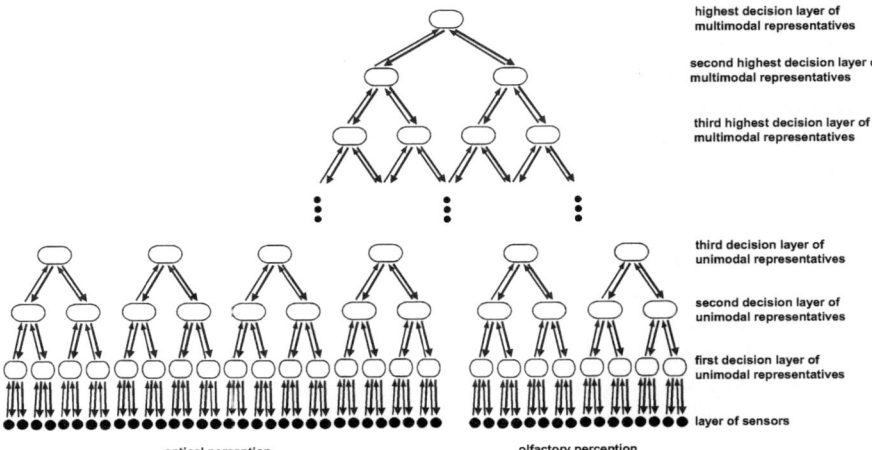

Fig. 6.19 Formation of invariant representatives in the different functional layers, modified after Hawkins (2004, p. 123)

something etc. However, they always receive the patterns (information) that they actually expect via feedback from "above". These amounts of information from the upper layers in the direction of the sensors are much greater than the information that reaches back into the upper area of layer L2 (see the different arrow thicknesses in Fig. 6.19).[41]

This means that if an upper function row expects a human face, it communicates this down to the lower row, which in turn breaks that pattern down into smaller patterns and communicates them down, and so on through all levels, until the lowest function receives an input that predicts what it should actually receive from the sensors. A search is then initiated, and the input is communicated back up. Figure 6.19 shows this in a very symmetrical way. In reality, the physiological picture of the sensor and neuronal system does not look like this, but certainly shows enormous asymmetries that are compensated by the information processing. People who wear glasses are familiar with this compensation effect when switching from one pair of glasses to another. The "eye" often has to "get used" to the new glasses. However, it is not really the eye that adapts, but the information processing of layers L1 and L2.

Figure 6.19 shows something else that Jeff Hawkins points out several times in his explanations: The patterns of the different senses are always processed in the same way, both in the unimodal as well as in the multimodal domain. Visual images are processed in the same way as acoustic or olfactory images.

---

[41] Wolf Singer writes (freely quoting) (Singer 2006, p. 25): "There is hardly a forward connection that is not paralleled by a quantitatively more powerful backward connection", thus confirming Jeff Hawkins' hypothesis.

This theory of the hierarchical formation of invariant representatives of expected objects solves three central questions that have long preoccupied me: Why are the images of my inner world invariant and do not change with the saccades of my optical system? Why do I recognize a friend no matter what angle I look at him from? Why do I recognize the friend even after 40 years? However, it is still an open question in SiMA which functions need to be developed for layer L1 (neuronal layer) and which for layer L2 (neuronal symbolization) for these tasks.

Another open question in SiMA is: What do people learn with regard to perception and what is genetically predisposed? The answer is crucial for a model of layers L2 and L1. All this can be modeled and tested by simulations and worked out experimentally, but from a computer engineering point of view only with the help of neurological and psychological experts, and certainly not without them. Anything else would be hybris.

It is important to point out that the complexity of the functions of layer L2 alone clearly shows that the forced direct fusion of neuronal layer L1 with layer L3 (psychic apparatus) or even directly with consciousness must necessarily fail. Many scientists, especially in the Blue Brain Project (2005), have great hopes in this regard, and even respected personalities such as John C. Eccles (1994, p. 122) have repeatedly tried to go down this path. In my opinion, this is also hybris or ignorance of the natural scientific facts. In discussions, I often reduce the explanation to a simple example: What sense does it make to connect Microsoft Word functions directly to transistor functions? Such thoughtful constructions of connections are meaningless in computer engineering, and even more so in brain research. One answer can only be a layer model such as that of SiMA, developed on the basis of the extended Mealy theory.

6.3.7   Data Structures: Data Model

It is certainly not the intention of this book to present technical program details. Such details can be found in the dissertations and other scientific publications listed in the bibliography. However, I have pointed out several times that at least some principles of the data structures, on which SiMA is based, need to be explained in order to better understand the problems of simulation in detail.

To briefly recapitulate: in the illustration in Fig. 6.4, particular emphasis was placed on the distinction between the three functional layers, albeit in the context of the progression of information between them, which is represented by arrows representing the transitions between layers. We will now take a closer look at how to describe the progression of information within each layer. Figure 6.20 shows this in a highly abstracted form. The physical and chemical effects on the sensors cause the connected neuron nuclei to generate electrical impulses. The distances between the impulses contain the information that needs to be transmitted. Starting from the transition i2 from layer L1 to layer L2, they are considered as symbols, which are used to describe the information processing in layers L2 and L3. The term symbol is

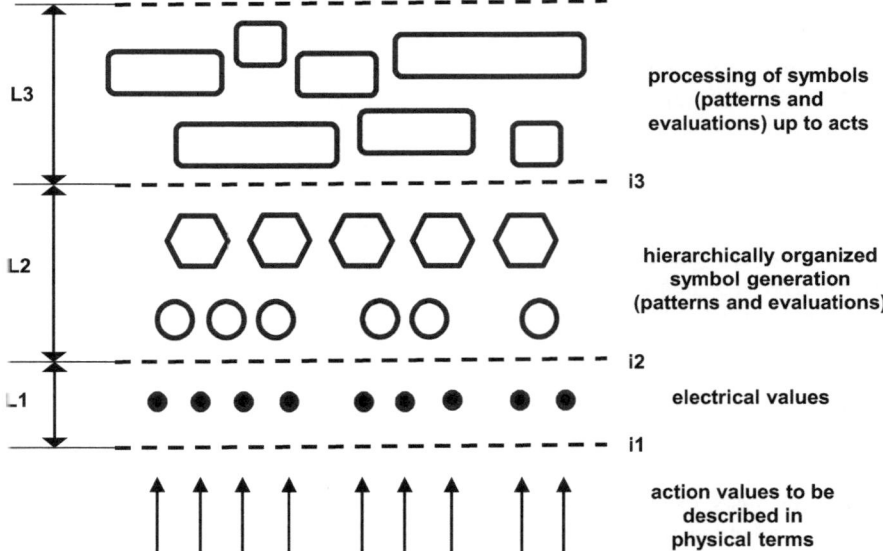

Fig. 6.20 Progression of information

meant in a mathematical sense. SiMA distinguishes between two types of symbols. On the one hand, there are patterns that represent objects (e.g. coming from the eye: e.g. edges or colors) or other physical perceptions, such as the state of homeostasis in the blood or the temperature of the body. They can also be patterns that form in the L2 symbolic layer based on the perceptions of the various sensory organs. All these symbols are patterns processed, transmitted and stored in both layers L2 and L3.

On the other hand, there are symbols that are not to be seen as patterns, but as quantities of valuations of something, i.e. quotas of affect, emotions, or feelings.

As described in the previous chapter, patterns of perceptions are composed hierarchically (see Figs. 6.17 and 6.18). When we see the pattern *hand*, this pattern is hierarchically linked to the patterns *fingernails*, *knuckles*, *wrinkles* and so on. The pattern *hand* is in turn part of the pattern *arm*. I have already explained that people also store *movements* as patterns. I mentioned my wife's particular way of moving, which is why I can recognize her when I see her from behind in the dark. However, humans are not able to store longer-lasting processes as individual patterns of movement. If a person wants to walk from their office room to the room where the coffee machine is located, they have to remember the individual locations—i.e. the corresponding patterns of the individual locations—on their way there one after the other. In SiMA we call these processes *acts*. The act of getting coffee cannot be represented as a whole, that is, as a pattern. It is important to understand that, according to Fig. 6.19, we only store patterns that represent the external world in an *abstract* way. They are a *model* of the external world.

To describe this mathematically and formally, computer engineers can use the mathematical formalisms of scalars and vectors. Scalars are single values, such as a

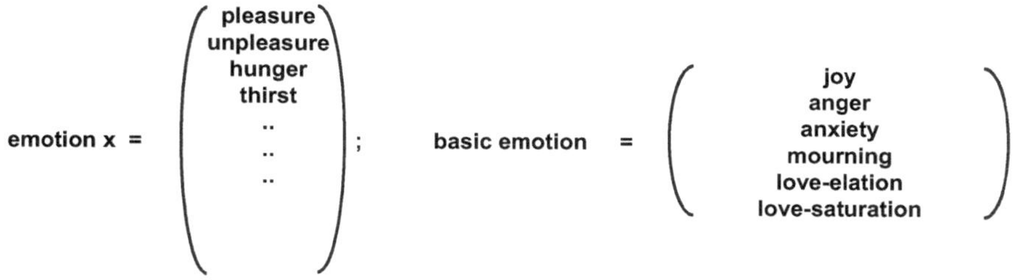

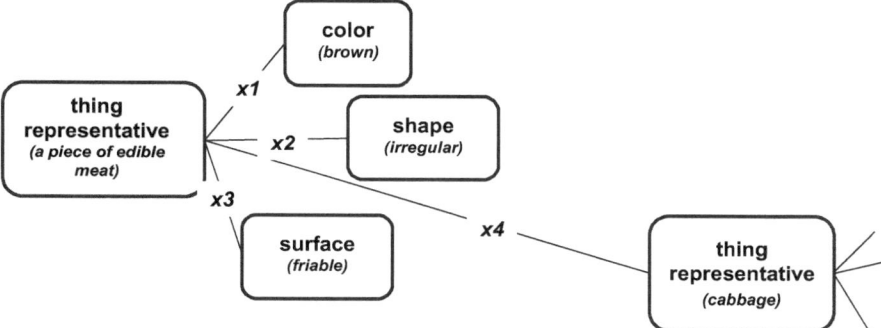

Fig. 6.21 Vectorial, symbolic representative of valuations

Fig. 6.22 A simplified representative list of thing representatives such as a piece of edible meat *(x1 to x4 are valuation variables (values between 0 and 1) such as quotas of affect or emotions)*

numeric value or a symbol. Vectors are made up of one or more scalars. For example, if we look at the emotion (= vector), i.e. the valuation variable of thing representatives, it is composed of scalars (in the case of an emotion: different quotas of affect), as Fig. 6.21 shows. The same is true for basic emotions. If you want to describe the pattern of an object, e.g. a coffee cup, this pattern (symbol) forms a network, i.e. it is connected with many other patterns such as shapes, materials, etc., and each of these patterns has a valuation variable.

This leads to the topic of networks of thing and word representatives. In SiMA, the crucial difference between the primary and secondary processes is that in the primary process, the representatives of a thing do not yet have a word associated with them, which means that they are necessarily unconscious. They can become conscious when they are assigned a word during the transition to the secondary process. In SiMA we therefore distinguish between patterns that are present as *thing repre-sentati*ves, i.e. *all* patterns in the primary process, and *word representatives* that are present *only* in the secondary process.

Each word representative is linked to at least one object representative. In addition, of course, all representatives are linked to properties and other representatives (Fig. 6.22). And all of them are assigned a valuation (a quota of affect or even an emotion). This means that *every* thing- and word-representative is valuated, regardless of whether it takes the property unconscious, preconscious or conscious. Therefore, if someone tries to judge something in a "neutral" way, as is expected of scientists, judges, and reviewers, this must be considered—to use a cautious phrase—only "relatively" possible. I think almost everyone is aware that it is at least difficult to decide and judge "objectively", "impartially" or "neutrally" on the basis of one's own feelings. It can hardly happen spontaneously. It requires a great deal of mental effort.

I think anyone who has dealt with such data structures understands that the complexity of the networked information system of the two layers L2 and L3 is huge. Incredibly huge! It means that this complex system has to be triggered every time you just want to get a cup of coffee, for example.

There are two other aspects that need to be mentioned. First, emotions, like basic and extended emotions, are derived from quotas of affect (Fig. 6.8). Therefore, they can only be described using more elaborate formalisms (Fig. 6.21). However, feeling—which is only present in the secondary process, i.e. the area that contains partially conscious information—cannot be a vector, because we only feel *one* feeling at *any* time, not different components of it. The feeling can indeed change quickly, but it is always just *one* feeling at any given moment. This means that it is calculated from the emotion vector and is also a scalar like the quotas of affect.

Second, due to the highly complex structure of the symbol system, it is perhaps easy to understand why the basic and extended emotions have not yet played a role in previous simulations. The fact that all vectors have to be included in the simulation package is only *one* aspect. I consider another aspect to be much more time-consuming: The tests to be developed by psychoanalysts in order to work out this differentiation of the quotas of affect and emotions in a satisfactory way and, above all, to be able to evaluate their underlying theory, will be very labor-intensive. In order to draw reliable conclusions, extensive statistical methods and tools will have to be employed.

6.3.8   Memory: Storage

Marianne Leuzinger-Bohleber has already been quoted several times in this book, especially in the foreword to Chap. 3. She emphasizes—similar to Mark Solms (Solms and Turnbull 2002, p. 18) and Jaak Panksepp (1998, p. 20)—that the Ψ-organ is not to compare with a computer and explains that even the principles of memory are completely different (Leuzinger-Bohleber 2008). A humans storage (memory) works associatively, a computers storage is address oriented. We know that Marianne Leuzinger-Bohleber, Mark Solms and Jaak Panksepp are not experts in computer technology, especially not scientists in this field. They are not really familiar with computer-specific technical terms and principles. However, there are experts in my field who put it even more dramatically. I pick out the sentence from Bethge and Dworschak (2011, p. 132) (freely quoting): "The brain is fundamentally different from a computer." Jeff Hawkins, the inventor of the Palm (PDA, palm history) an outstanding computer specialist whom I often quote, wrote: "Brains are pattern machines." (Hawkins 2004, pp. 61–62). He also emphasizes that Ψ-organs are not computers, which he explains even more concretely elsewhere, for example in Hawkins (2004, p. 39).

Such bold claims, without a clear definition of the term computer, without concrete scientific delimitations or comparisons, should be questioned in principle, because it is often based on false prejudices. Formulated in this way, they are worthless—like the above statement by Jeff Hawkins, who is otherwise an exceptional scientist.[42] They evoke emotions that can lead us via associations to wrong conclusions. I would like to counter this.

How can the contradiction between their statements and my judgement that the Ψ-organ is a *(biological) computer* be explained? The people mentioned use the term computer in the sense of common parlance, while I, as a scientist and professor of computer technology, use it in the sense of the theory of computer systems. In common parlance, the term computer refers to a device made of sheet metal, silicon, and a lot of plastic, perhaps even the von Neumann architecture, an architecture that is still used in computer technology today. This parlance already overlooks the purely mechanical and, above all, the long-established analog calculator. Granted, at that time these devices were primarily called analog calculators, but at the same time we also spoke of digital calculators (and not digital computers—especially in the German speaking countries). For a long time, we all had difficulty using the term *computer* in the German-speaking world. My first professorship in Bielefeld was also designated as d*ata processing*. In Vienna, too, our institute was renamed from the *Institute for Data Processing* to the *Institute for Computer Technology* only after my appointment. Many people were suspicious of the word *computer*. It was not much different in France. There, people spent a long time looking for a purely French term.

We should also consider the ideas of Alan M. Turing and his principles of computation. It is important to remember that in science[43] it was assumed very early on that the various approaches of Alan Turing, Konrad Zuse and John von Neumann or Heinz Zemanek should be seen as the first steps. Others had to follow and others still have to follow. Today, for example, we often hear about the quantum computer, which—at least in its core—will have little in common with our current machines. Computers based on neural networks are also conceivable, but the necessary mathematics of threshold logic is still not nearly as well understood as that of binary logic.

---

[42] The question must be asked: Why do the terms *computer* and *pattern machine* have to be a contradiction at all? According to the clear definition of the computer in SiMA, they are not. And if I interpret Jeff Hawkins' thoughts in his book correctly, I understand what he is getting at: If computers are to become "intelligent", they must have the functional structure of a human Ψ-organ.

[43] Note that Jeff Hawkins worked as a computer technician, but never worked as a computer scientist. He did, however, go on to do research in the field of neuroscience in the field of perception.

What am I getting at? Terms often change their meaning over time, as does their emotional valuation. This is what happened to the term *computer*. But are the contradictions only due to confusion of terms?

For all these reasons, I consider the *most general and abstract definition of the computer* to be useful, as I explain in Sect. 3.1: *The computer is an (abstract) system that can process, store and communicate (large amounts of) data.* This definition does not specify a physical realization. The logical conclusion is that every natural Ψ-organ is therefore necessarily a biological computer. At the same time, as Jeff Hawkins brilliantly and crucially formulated it, we can call the Ψ-organ a *pattern machine, because* the L2 and L3 layers exclusively process valuated patterns (described by symbols).

Let me return to the main topic, the storage of data (numbers, symbols, etc.). I do not want to talk about mechanical storage principles as well as data storage via quantum effects. According to the current state of science, I know of no evidence that speaks in favor of them playing any role in storage, processing or communication in the Ψ-organ.[44] However, research into the storage of information via neural and digital functions has progressed to the point where the two possible general storage principles discussed in Sect. 3.1 could be considered scientifically established. One is the principle of feedback, as shown in Fig. 3.16, and the other is the principle of synapse formation, which corresponds to the principle of the electronic ROM circuit. In the Ψ-organ, the neuronal feedback principle is used for the short-term storage of information. This is commonly referred to as short-term memory. Mark Solms does not like this and therefore speaks of working memory and gives specific values for it. Information is held in working memory for a maximum of a few seconds (Solms and Turnbull 2002, p. 143) until an inhibitory effect overrides the positive feedback and the information is lost. In electronics, this type of working storage, which operates according to the feedback principle, has the disadvantage that the information is completely lost when the supply voltage is switched off. Short-term memory is contrasted with long-term memory, the storage of information through the formation of synapses, which are only lost through their slow degradation (long-term inactivity) or destruction (Kandel 2006, p. 214).

Does this mean that in principle there is no difference between the memory of a human Ψ-organ and the storage of a technical computer? Are the people I mentioned above, such as Marianne Leuzinger-Bohleber, Mark Solms, Jaak Panksepp, Jeff Hawkins, etc., really wrong? Yes and no! Of course, there are association memories for technical computers of various designs. If you develop an association memory directly in hardware, it will be expensive. As it is not mass-produced. So, a detour is taken: Address-oriented memories (ROMs and RAMs) are used in the lowest access layer, and a function (an additional layer) for association coding is placed on top.

---

[44] As I mentioned earlier, some people are trying to consider the quantum effect as a possible principle of information processing or storage in the Ψ-organ. I see one critical point that needs to be raised in order to be able to follow this idea: How can the functional layer of quanta in the Ψ-organ be related to electrical functions? Until a scientific model is available, this path should be viewed critically.

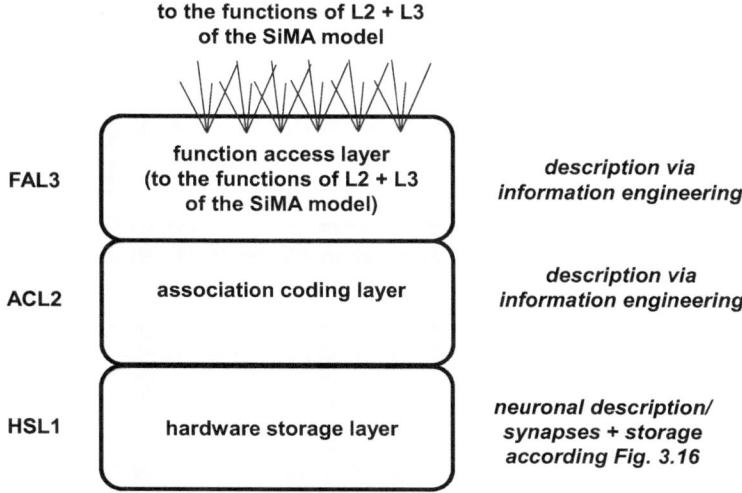

**to the functions of L2 + L3
of the SiMA model**

FAL3   function access layer
(to the functions of L2 + L3
of the SiMA model)

*description via
information engineering*

ACL2   association coding layer

*description via
information engineering*

HSL1   hardware storage layer

*neuronal description/
synapses + storage
according Fig. 3.16*

Fig. 6.23 Functional layer model for storing data in the Ψ-organ

The result is a multilayer functional model, as shown schematically in Fig. 6.23. The lowest layer is the *hardware storage layer* (*HSL 1*). It is hardware and mass produced. The layer above it is the adaptation layer; we can call it the *association coding layer* (*ACL2*). It can be implemented in hardware or in a combination of hardware and software. In our SiMA model, we also need a *function access layer* (*FAL 3*) for the individual functions of the SiMA model in layers L2 and L3, as shown in Fig. 6.14. In any case, it is implemented in software, since all functions of the SiMA model in layer L3 are currently executed in software.

Because it is important for the functioning of the psychic apparatus, I would like to point out once again that all patterns, such as thing or word representatives, are always stored with a valuation. Thus, when the psychic apparatus accesses memories through the three layers of memory shown in Fig. 6.23, it always receives a thing or a word representative that is associated with a quota of affect, an emotion, and/or a feeling. And all storage and retrieval can only take place through these three layers.

An associative memory as shown in Fig. 6.23 can explain interesting phenomena of human memory for which there was previously no scientific evidence and which can now be evaluated by SiMA. Let me give you an example. My mother's dying process was long. She lost more and more of her memories, but instead new memories emerged that we children had never heard before. What was going on in the increasingly destroyed Ψ-organ? We know that all organs renew themselves in the course of a lifetime. On the one hand, when synapses die, the information they contain disappears. On the other hand, the psychic apparatus is constantly working with both unconscious and conscious content (including repressed and primarily repressed content), which is mixed with new content and, above all, its valuation. Think of dreams, where memories are constantly brought to surface. This modifies the memories, they receive a different valuation, but are stored anew (modified) by their activation. This may be the reason why we can remember so far back into our childhood. And it may also be the reason why the memories of older married couples often completely contradict each other. Let us go back to a person whose Ψ-organ is slowly being destroyed in the process of dying. The synapses are destroyed, and with them the memories. Other memories in the same chain of associations, which previously had a lower rating than the now destroyed memories, are indirectly revalued by the disappearance of the higher-rated memories and come to light. This also means that non-causal connections are modified and networks of thing and word representatives are created that can no longer be traced back.

These are hypotheses that I have taken from neurology and psychoanalysis. The SiMA model of memory now allows us to simulate and evaluate them experimentally. However, you have to be aware that the effort is not minimal and, as with SiMA, can only be done efficiently by an expert team of neurologists, psychoanalysts, computer engineers and computer scientists.

And one additional remark. The reader may have noticed that I have again omitted *learning*, although learning requires *information storage*. Both functionalities are closely linked. Almost all functions of the psychic apparatus (layer L3) contain the functionalities of learning and information storage. However, we have not yet succeeded in breaking down layer L3 (psychic apparatus) of the three-layer model (Fig. 6.3) into its individual functions (Fig. 6.14) for the memory model (Fig. 6.23). This work still needs to be done in connection with the functionality of learning.

I still have to deal with two contradictions about memory that I find surprising that they are rarely discussed. The first contradiction: We now know that the neuronal system is an organ like any other, which means that it renews itself over time, probably over a time span of about 12–15 years. Whether this value is correct or not remains to be seen. The interesting question is: What happens to our memories? Are the molecules in the synapses replaced 1:1 by other molecules? My previous understanding was that whole cells are replaced. In this case their synapses also disappear, and with them the memories stored in them. Yet I can still remember more than 15 years ago. How is this possible? My hypothesis, which is in line with the theory of psychoanalysis, is that through dreams and constantly activated memories (especially in the primary process), memories are, unconsciously and consciously, are restored with new valuations of quotas of affects, emotions, and feelings and thus remain intact, albeit modified, over the course of a lifetime. Only the details that are no longer remembered disappear forever.

The second contradiction is more complicated for me, but again I have a hypothesis. As I mentioned earlier, Mark Solms and others put short-term memory at a maximum of 2–3 s. Whether this value is exactly correct is again irrelevant. In any case, one can imagine that in this time no synapses can be to store the information in such a way that it becomes accessible as long-term memory. So, how does learning work? Although the problem was raised in SiMA, it has never been investigated in detail (Fittner 2021). My hypothesis, which again is compatible with the psychoanalytic model, is that according to Jeff Hawkins, you can only

Fig. 6.24 Model of
information flow between
functions to explain
different types of possible
aphasia [modified from
Freud (2001, p. 45)]

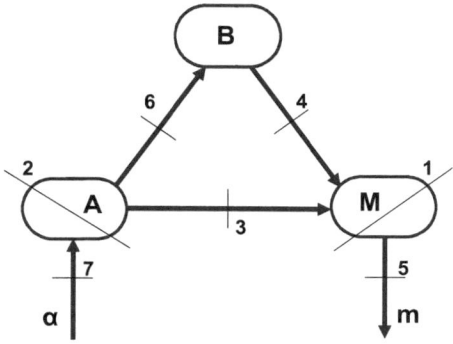

perceive something if you can associate something familiar with what you see (see
Fig. 6.19 and the accompanying explanations). This process, in turn, activates other
objects. One has to determine whether what is perceived is the same, similar, or
different. If it is different, what object is it similar to? This means that the perceived
object that triggers a high valuation (quota of affect, emotion) is repeatedly dealt
with in the primary process, not only in the short term, but also in the long term. It
triggers a chain reaction of short-term storage, which can ultimately lead to the
formation of synapses, whereby it is "shifted" into long-term memory.

However, these are hypotheses that I have put forward and that need to be
examined more closely from a neurological and psychoanalytical point of view,
since they are essential for a simulation of the Ψ-organ. So far, we have had to make
do with assumptions.

6.4   Distributed Systems: "Zur Auffassung der Aphasien"

Yes, in the title of this chapter I make a direct reference to Sigmund Freud's early
publication in which he subjected the theory of localization of aphasia[45] to rigorous
scrutiny and earned himself considerable hostility: "Zur Auffassung der Aphasien—
Eine kritische Studie" (Freud 2001) (the title loosely translated: On the Conception
of Aphasias—A Critical Study). He espoused a perspective that differed from the
mainstream in his field. His study left a profound impression on me.

> The monography shows that Sigmund Freud back then already thought functionally, in a
> manner similar to that of contemporary computer engineers, over fifty years before the
> fathers of computer engineering, including Konrad Zuse, Claude E. Shannon, Georg
> H. Mealy, and others should develop technical information theories.

---

[45] Aphasia is a disorder of the speech system. The lines numbered 1–7 in Fig. 6.24 thus represent the
potential disruptions to the speech system in humans.

He was ahead of his time when he argued against Carl Wernicke (Freud 2001, pp. 7, 45): It is not possible to localize psychic functions neurologically/physiologically. They are functions of the Ψ-organ, which processes information, and should be considered independently of their physiological locations within the Ψ-organ. This is a way of thinking that is applied in the Mealy automaton and the ISO/OSI model. In the first step, both models are independent of the physically, chemically, or physiologically described functions. I would like to explain what needs to be analyzed in the next step using Fig. 6.24.

Sigmund Freud's critical study also indicates that words, sentences, and their content are associated with conscious, valuated ideas.

> Consequently, speech recognition systems or translation software will only reach the same performance level of humans when they have been provided with a consciousness—whatever form it may have.

This issue must therefore be addressed in AI, even if it is challenging for natural scientists. However, natural scientists have frequently been confronted with the problem that their ideas, that were constrained by their zeitgeist, were not sufficient to address the needs of their time. Instead of elegantly circumventing the issues, one has to tackle them. For instance, consider the calculation of the movement of celestial bodies, Charles Darwin's theory of evolution, psychoanalysis or the theory of relativity. Initially, there were vehement objections and arguments to reject these new ideas. Traditional ways of thinking contradicted such seemingly absurd views.

Let us return to the critical study by Sigmund Freud. A central figure[46] of that study is shown in a, according to information engineering presentations, slightly modified form (Freud 2001, Fig. 3, p. 45). α is the sensory input variable, m the motor output variable. The numbers 1–7 denote the different possible aphasias. A, B and M are functions in which acoustic information, acoustic signals and even words are processed. M is the motor-acoustic output function, which can be expressed with the SiMA model of Fig. 6.14 as a part of the functions F30, F31, and F32. A is the input function of the acoustic information channel, which can be expressed with Fig. 6.14, as a part of functions F10, F11, and F14. Functions A and M can be neurally localized in the Ψ-organ, but this is no longer possible with function B. According to computer engineering, it can no longer be assigned to a specific location in terms of hardware. The task of function B is performed by many neurological units (hardware). This is also true of all functions of the psychic apparatus (layer L3), and presumably of layer L2. If one wishes to move the biceps, the functions F31 (L2) and F32 (L1) of Fig. 6.14 are activated. It is evident that some of the neurons on which they are based can be recorded electrically. However, the question remains as to whether the neurons of function F30 (L3) can still be recorded. In computer engineering, the neurons of F32 and F31 are referred to as the output drivers of a processor. The visualization of such information in electronics is therefore a self-evident matter. The same applies to the input functions of the

---

[46] An error has crept into the edition I am quoting (Freud 2001). The designation "Fig. 1" appears twice. The second figure referred to as "Fig. 1" is actually the third. I therefore quote it as Fig. 3.

Fig. 6.25 Decentralization of a computer by means of additional units such as a number cruncher, an address driver, etc.

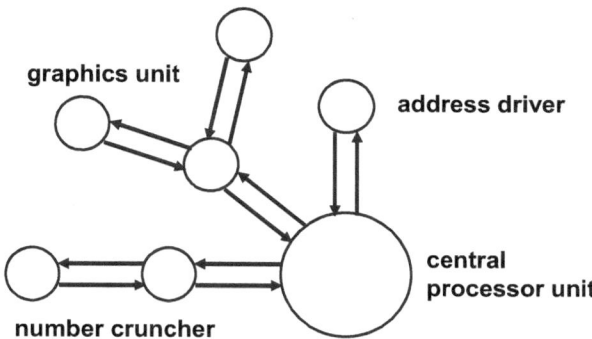

graphics unit

address driver

central processor unit

number cruncher

psychic apparatus (layer L3). For all the functions *after* these input functions or *before* the output functions, this possible neuronal localization no longer applies. In computer engineering, a function that can no longer be assigned locally to a specific piece of hardware is expressed as follows: *Function xy is distributed across different parts of the hardware.*

Let me explain this illustration in a slightly different way.

When a new technical communication tool is developed, its performance must first be defined. This will tell you what functions and layers are needed, as shown in Fig. 5.8, and you can then estimate which of these should be developed via hardware and which via software. Assigning, especially after the fact, which software function has which hardware reference is generally not relevant.[47] This means that today's computers are *distributed systems* in two ways. By *distributed system*, I mean a system in which there are multiple functions in which information is processed. Everyone is familiar with *distributed software systems*. Just look at the different apps (applications) on your smartphone. Each app represents such a function, where information is processed in a certain way. The different apps together form a distributed system.

Separately, high performance computers include "smart" add-on units (hardware functions) that support the central processors for fast information processing (Fig. 6.25). They are assigned complex tasks for which they are specially developed, making them much more efficient than the central processors. Examples for this are number crunchers, addressing or graphics units. More powerful computers today therefore generally consist of several hardware units, i.e. a distributed hardware system that is in principle independent of the distribution of software functions.

---

[47]For better understanding, I repeat myself: The psychic apparatus does *not* correspond to software. The psychic apparatus is a *function*. Software is the formal tool with which such functions of the information system can be described in detail. However, we do not know the software that can be used to describe the function of the Ψ-organ. According to the SiMA model, we only know the sub-functions (Fig. 6.14) that we can map following the tools and procedures of information engineering.

Going one step further and considering supercomputers such as the Summit with its 27,648 graphics processing units and a possible performance of 122 petaFLOPS, one is faced with a sea of "smart" components (processors, graphics units, etc.). The dynamic allocation of tasks is handled by special functional units in the computer.

What am I getting at? The principle of functional hardware distribution (= different neuronal areas) in the Ψ-organ can also be found in computer technology. In the same way, certain tasks in the Ψ-organ as well as in the computer can be localized in terms of hardware, while others cannot. They are then "smeared" over large areas. And I think this is what Sigmund Freud wanted to point out in his work "Zur Auffassung der Aphasien". Conscious thinking, feelings, etc. cannot be localized neurologically because the functions responsible for them are distributed over large parts of the brain. However, Carl Wernicke, who was considered a "beacon" of brain research at the time, disagreed, as did most scientists in the field (Freud 2001, p. 7).

I must object to one point in the introduction of new edition of the monography "Zur Auffassung der Aphasien", because it can lead all too quickly down the wrong path and into dead ends, where some things must seem mystical to some people. Wolfgang Leuschner has written an excellent introduction to the new edition of Sigmund Freud's monograph, and he explains very well why it was necessary. However, at one point (Freud 2001, p. 17) he literally adopts a poor formulation of Sigmund Freud (2001, p. 98) (freely quoting): "The psychic, then, is a parallel process of the physiological." This is incorrectly formulated. Physics and thus physiology is a tool to describe material functions in a certain way—physically/physiologically—while the psychic apparatu"s is a function that can be described efficiently in terms of information engineering. It is one and the same model, one part described physically and the other described in terms of information engineering. This is not just nitpicking about jargon. There is no parallel process. As a student, I never understood how physics, chemistry, and information engineering fit together. Are there parallel functions like physical, chemical, and information? Shouldn't there be a unified formula for all three? A world formula, so to speak? Are there other such fields? If it is clearly expressed that these fields are only tools for describing each other, then the extended theory of Georg H. Mealy or that of the ISO/OSI model becomes understandable. And the interrelationship between the fields loses its confusing character, and thus they can be thought together.

6.5   The Self

Consciousness necessitates a self. That is, I must have an image of myself as an individual. The self forms an image of me for me.

For a long time, psychoanalysis defined primarily the *functions* of the psychic apparatus that makes people act and react. The *function of the self* was largely ignored. However, in SiMA, over time, the question crystallized: *How does a human being perceive themselves*? It is difficult to determine the self of a human being. For a long time, this question was elegantly evaded, until Klaus Doblhammer could firmly establish it as a funded project in SiMA (Doblhammer 2013). We quickly realized that this topic is quite challenging. Although Sigmund Freud already saw parts of the self in the ego function he defined (Hartmann 1950), the topic has always been very controversial in psychoanalysis.

So, what is the self? How can it be defined? If researchers want to simulate the model of the Ψ-organ, they must address this question. An individual is characterized by its *self representatives*. If an individual is to become capable of learning, this presupposes a function of the *self* that forms and works with the *self representatives* of its own body and its own psychic apparatus. And very importantly, *the self is the prerequisite for an individual's ability to become conscious*. Individuals must have a presentation of themselves if they are to build a relationship with something else.

This leads to the question: What characterizes conscious human beings? Certainly, the ability to communicate and understand each other. The case studies in SiMA must therefore involve complex, interacting individuals. These individuals differ not only in their memories, but also in their individual self representatives. This is one of the reasons why the agents in SiMA have so far only been allowed to live for a few minutes in the experiments—just to simulate a short episode (called *acts* in SiMA), which allows the model to be evaluated over a limited number of steps. If one tries to significantly increase the number of memories and to vary the self representatives according to external situations and internal states, an interpretation of the behavioral processes and thus an evaluation of the experiments becomes extremely difficult or even impossible. However, this is also the reason why Klaus Doblhammer finally demanded a detailed study of the self. Based on this, the principle of *learning* of the various functions of the psychic apparatus can be developed in order to use this knowledge to gradually approach the simulation of *consciousness*. I will come back to this in the next chapter.

Back to the study of the self. Some psychoanalysts have dealt with this topic. As already mentioned, they arrived at contradictory conclusions. SiMA, however, requires a clear axiomatic definition and a functional explanation. A behavioral description is not sufficient. From the point of view of computer engineering, the self is not too difficult to work out, but from the point of view of psychoanalysts it is all the more difficult. This is quite understandable. To explain this, let us imagine a future *learning* automaton such as an intelligent robot that we hope to create in the (distant) future, and compare it with a human being. This robot uses a large variety of sensors to continuously determine its status. Are its mechanical and information functions impaired in any way? Does it still have enough physical energy for the next tasks? Depending on where it is used, it must also be able to adapt its behavior accordingly. The dependencies can be: tools to be carried and operated, weights to be lifted and balanced, path and road conditions, obstacles, humidity, temperature, even the weather in general, and differences in lighting that it must constantly assess for itself. A burning house requires different behavior than working in a snowy wasteland. How must it adapt to all the different situations and circumstances? How does its body react to the changing external conditions? How elegantly (or less elegantly) can it react to external conditions, respond to external processes, or even, if possible, control them? The many differences in possible processes cannot be fully predicted, which is why in SiMA we speak of complex processes. The robot will have to approach them in a *learning* way to find out which of its possible behaviors are the most efficient? Robots of the future will have to use their own perception to gain ever better knowledge of themselves and their complex behavior in their often highly variable environment. This includes functions that develop this knowledge. In SiMA, we call the superset of these functions the *self,* whose sub-functions represent partial functions of the function of Fig. 6.14. These subfunctions develop together *self representatives*, i.e. the patterns (data) that characterize the individual.[48] They are continuously memorized and are available to the individual to know who they are and what state there are in. Based on this principle, the control systems of robots can be functionally constructed in a similar way, but they must adapt efficiently and individually to their tasks. This requires learning phases.

What am I getting at? When we engineers design the robot of the future, we can precisely determine the individual subfunctions of the functions in Fig. 6.14 that make up the self and, most importantly, which we can simulate. Yes, even more, *we can integrate monitors*[49] *into the SiMA simulation model as well as into the robots, that observe and monitor all of their states and self representatives.*

In this thought experiment, let us switch to the human being. Why are *self representatives*, i.e. the self, so difficult to grasp in psychoanalysis? We cannot integrate monitors into the functions (Fig. 6.14) of humans. Therefore, it is not possible to measure self representatives objectively. They are exclusively subjective and will remain subjective. They are not perceivable to others. As Sigmund Freud explained in his scientific work *Zur Auffassung der Aphasien* (see previous chapter), they cannot be grasped neurologically, but as Antonio R. Damasio clearly stated (see Sect. 4.3), the natural scientist has no choice but to deal with them. And this is also my incentive to make progress in this area.

Let us get back to the subject of this subchapter. As Klaus Doblhammer explained (Doblhammer 2013, p. 7), he did find different conceptions of the self in the literature:

- the temporally stable self as opposed to the fluctuating self,
- the self as an object of perception versus the self as an indivisible, immediate experience,
- the bodily self, the psychic self,
- the true self as opposed to the false self,
- the public self, the private self, or the autonomous self
- etc.

---

[48]I repeat: It must not be forgotten that an individual is determined not only by their self, but also by their memories.

[49]Monitoring is used in the simulation units of the SiMA model that log all important processes. The software units used for this purpose are called monitors.

It can be seen that some terms refer to a certain behavior. It has not been possible to provide an axiomatic definition. Therefore, such terms are not suitable for a natural scientific approach as in SiMA, since, as explained at the beginning, the goal cannot be the simulation of behavior, but must be the simulation of functions.[50] And such a simulation must be developed with psychoanalysts, the specific experts of the psychic apparatus. Based on such considerations, the team of the Project Self in SiMA (Doblhammer 2013, p. 2) has primarily followed the explanations of Joseph Sandler (Sandler and Sandler 1998) and Otto F. Kernberg (1995). They use Freudian conceptions, methods and terms and thus come closest to the specifications of the SiMA project. From their descriptions it can be deduced that people arrive at *self representatives* through perceptions of the external world and the associations of their memory. Consider the example of a human individual who sees a cup. Through an intact reality and body check, it is identified as a part that does not belong to their body, i.e. as an *object representative* that is distinct from their body. If, on contrast, the person sees their hand, it is part of the *body representative* and thus a *self representative*. The same applies to body movements, the way we speak, or our reactions to events. For example, if you take the wrong medication, or if medication is not taken correctly, your body parts can suddenly behave differently. Your proprioception changes. Your own body parts can become foreign to you. In other words: The perceptions no longer correspond to the remembered *self representatives*.

Now the question remains: How can the parts of the functions of the *self* be integrated into Fig. 6.14? This mapping is shown in Fig. 6.26. The sensory information (1) and (2) of the perception track (PT) is transformed via the neuronal system layer (L1) and neuronal symbolization layer (L2) into function F14 (ExPe: external perception). Fantasies are added via function F47 (transformation to the primary process). All this information is linked by memory traces (that is why F14 is represented by a rectangle), which are valuated (cathected) by quotas of affect. The quotas of affect come from the *applied drive track*, specifically from function F57 (generation of drive representatives). The result are the thing representatives at the output of F14.

These thing representatives are linked with word representative during the transition to the secondary process and arrive at function F69 (extraction of possible goals). There, two tasks are addressed, which I show in Fig. 6.26 in a simplified form via two subfunctions of function F69: F69a and F69b. Subfunction F69a performs a reality check, while F69b, which follows subfunction F69a, performs a body check. They check whether the perceived objects belong to one's own body or not. The reality check examines the valuations (psychoanalytically formulated: cathexis) of the representatives. If the valuations of the sensory information have a significantly

---

[50]Remember that scientific thinking assumes a functional unit that produces a behavior. The mere description of the behavior of an object remains incomplete, because without the knowledge of the description of the functional unit of the object, experiments cannot be carried out to their full extent. And experiments are a prerequisite for the natural scientific approach.

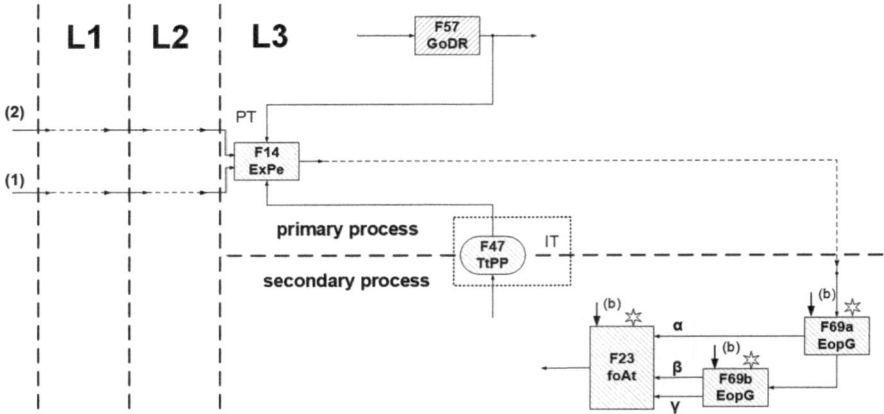

Fig. 6.26 Integration of the sub-function components of the self into the functions of Fig. 6.14 L1, L2 + L3: functional layers of the SiMA model; IT: imagination track; PT: perception track; α: mentation representatives; β: object representatives that are part of the own body (= body representatives); γ: object representatives that are not part of the body; for all other designations see Fig. 6.14

higher weight than the associations and fantasies, they are perceived as external reality. If the fantasies and associations predominate, these are to be regarded as *mentation representatives*, i.e. as representatives that are consciously perceived as pure phantasies and further utilized accordingly in subsequent functions. They bypass function F69b. If, on the contrary, they are regarded as objects of reality, subfunction F69b checks whether they can be assigned to one's own body, in which case they are identified as body representatives. If they have nothing to do with one's own body, they are generally identified as *object representatives*. These decisions take into account information from proprioceptive and pain sensors.

The self representatives are determined by the *mentation* representatives and the body representatives (in Fig. 6.26: α and β). For each individual, these are variables that are not static, but are constantly changing. When I am very hungry, I have a different self representative than after a sumptuous meal, which may even have been consumed with alcohol. And if I want to remember my self representative over many years, there is a legitimate question as to how far this is even possible. How does one consciously perceive changes in self perceptions, especially in detail? How can they be captured for SiMA experiments? The experts, the psychoanalysts, will have to answer precisely these questions. However, one thing is already clear from these considerations: How I see myself today is different from how I saw myself yester-day, 10 months ago, and especially 20 years ago. And: the *self* of a human individual is and will remain something subjective as long as we are not able to read out the self representatives of a human being—and I have no idea how that would be possible.

It should be noted that the self has not yet been simulated in SiMA and therefore has not been experimentally evaluated. So far, it has remained a model design without experimental hypothesis testing, which must be the next step from a natural scientific point of view.

6.6 Consciousness

The next research step towards the goal of modeling consciousness is the process of *learning*. Only through learning processes in functions of the Ψ-organ can sufficient knowledge about the self be gained, which is an essential basis of consciousness. Although the self has been developed in the SiMA project, it has not yet been evaluated experimentally. In this sense, learning should now also be approached. The first step would be to analyze how to define *learning* axiomatically, followed by a model adaptation according to Fig. 6.14, and finally experiments for evaluation. However, experimental evaluation of the modeling of the self has not yet been done. For this reason, I will not discuss learning further in this book. Another important reason is that all the functions of the Ψ-organ must be treated differently—as is the case with the concept of intelligence—because different learning goals must be defined for each individual function. This has been well demonstrated in Martin Fittner's dissertation (Fittner 2021). However, this can only be done in extensive projects in SiMA. The difficulty of this work should not be too great if the tasks of the individual functions of the Ψ-organ are broken down in sufficient detail (e.g. according to Dietrich et al. 2015) and a detailed description of the different learning functionalities is provided. Nevertheless, the work still needs to be done.

Therefore, I will now directly address the topic of *consciousness*. As far as the simulation of the Ψ-organ is concerned, it is still the last major missing element for me—apart from some details of the Ψ-organ. In my opinion, it is also the most difficult in many respects. With the help of psychoanalysis, as I hope I have been able to show, the terms unconscious, preconscious, and conscious can be defined elegantly and axiomatically based on natural science. It is not so easy with the topic of *consciousness*. Although we have discussed consciousness in the SiMA project, its translation into an axiomatic natural scientific system and its integration into the SiMA model is currently not possible. And this is mainly for what I see as a purely political (not scientific!) reason. The resistance of my colleagues in the natural scientific community is too strong. No graduate student, no Ph.D. candidate, and no member of the SiMA team has officially dared to tackle this problematic topic. And everyone who considered it, I strongly advised against it in their own interest, as I knew how a diploma examination or a doctoral defense in one of the technical faculties might turn out. I could certainly have recruited open-minded colleagues from abroad who would have written an expert opinion on how well the student was able to transfer a question from the humanities to a natural scientific problem. However, other colleagues—I call them "hardliners" from physics, electrical engineering or computer science, who had not yet dealt intensively with the interface between hardware and the information part—would have done everything they could to undermine these efforts during the examination and doctoral defense and to discredit the student. I have often experienced similar embarrassments simply because graduate students dared to relate the subject of *psychoanalysis* (a far less problematic term) to the natural sciences in their work. My experience has been that medical doctors are often more open to this topic, as the publications of neuroscientists such as Antonio R. Damasio, Oliver Sacks, Mark Solms, Yoram Yovell, or the Viennese Henriette Löffler-Stastka and Gerhard Wiest clearly show. I am convinced that the rejection of the attempt to approach the connection between consciousness and the natural sciences from a scientific perspective is based solely on a lack of understanding. Most of my colleagues are probably only familiar with philosophical or religious considerations of consciousness, whereas neurologists and psychiatrists have to deal with the natural sciences as a result of their studies on the one hand, but also have to approach psychological questions from a clinical perspective on the other. This removes the fear of the great unknown and the confrontation with everything that is associated with "our soul" or our subjectivity.

In the following, I will try to briefly summarize the results of our work in SiMA so that the reader gets at least a rough idea of what the next step in SiMA might look like. I hope that there will be scientific teams who will read this text and who will not be confronted by the "hardliners of physics" and their social pressure, and who will therefore be able to approach the topic of consciousness with an open mind and, above all, with fewer negative prejudice. For this reason, I consider the following thoughts on consciousness to be only a sketch and not yet sufficiently drawn out for a simulation. They are our initial ideas in SiMA. They need to be deepened in a *heterogeneous* team using *natural scientific methods*.

What are the requirements for consciousness? As I explained in Sect. 4.3, a distinction must be made between core and extended consciousness, with Mark Solms calling core consciousness *simple consciousness. Simple* and *extended consciousness* both require the representatives of the self, but use them for different purposes. Simple consciousness is concerned with the *now and here* (Damasio 1999, p. 16). It connects the information of the current external world with the current self representatives, which are processed and valuated through *quotas of affect* and *emotions* (Q, E in Fig. 6.27). The highly complex, extended consciousness requires that all *thing representatives* are linked to *word representatives*.

How can this graphic be interpreted with the knowledge developed so far? Why do I "see" my environment, "hear" music, or "smell" coffee? Human sensors (S) in Fig. 6.27 perceive physically described quantities outside and inside the body, whose information is encoded in layers L1 and L2 (see also Fig. 6.3) in the form of patterns described by information engineering. The patterns are thing representatives. The psychic apparatus adds associations of the remembered inner world to these patterns. All incoming information is functionally linked by various functions, which is visualized by the summarizing ensemble of functions f_a in Fig. 6.27. From this, the psychic apparatus generates the representatives of the current inner and external world. The following ensemble of functions f_b can use this information to determine the current self representatives. All representatives are valuated (cathected) by quotas of affect and/or emotions. However, this is not sufficient for me to see, hear, or smell anything. I only perceive the information via my sensors,

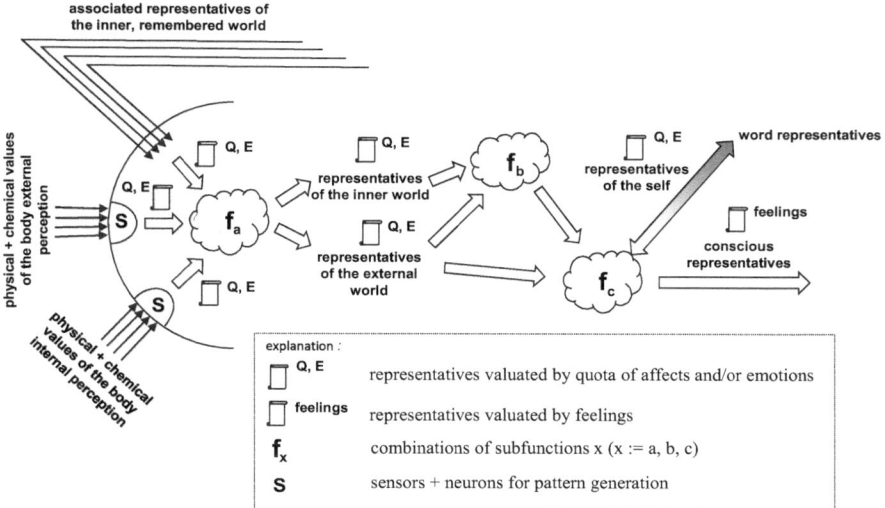

Fig. 6.27 Information flow to consciousness

which is valuated via quotas of affect. Only in the functional ensemble f_c does the assignment of words to the thing representatives take place. Then the last step, which is the most difficult to understand, can be achieved: seeing, hearing or smelling *what you expect to see, hear or smell* (Klein 2015). Here we should once again use Jeff Hawkins' illustration (Fig. 6.19). The eyes, ears, or nose (and of course this applies to all other perceptions in a similar way, as explained in Sect. 6.3.6) determine hierarchically from top to bottom what they want to see. And if the specification matches what is perceived, a conscious feeling of seeing, hearing or smelling arises. *We are experiencing consciousness. We are feeling.*

I was not entirely convinced by this line of thought, because the question remained: How can I see, smell, or hear something if I do not yet know what it is? But perhaps it is easier to understand if you consider that new objects like new images, sounds, smells or what ever that reach my eyes, ears and sense of smell initially only trigger the functional ensemble f_c to associate: images, sounds or smells respectively. *Only when there is a match between what is associated and what is perceived, feelings are generated.* Then you see, hear, or smell the corresponding thing. Consciously seeing, hearing, smelling, touching, etc. provide a *positive feeling* of *recognition*. I have been searching for this insight for a long time. I am fully aware that it is still a hypothesis, but everything I have described so far leaves me with only this solution. And I find it plausible. It should be evaluated using simulations based on the SiMA model.

The difficult topic of *consciousness* is closely related to another topic I would like to address here. It is much easier for me to deduce, even if it is all the more difficult to imagine. It is a topic that has only been dealt with intensively in religion and philosophy: *free will*. The three-layer SiMA model shows the Ψ-organ with all its essential functions, which makes decisions based on its memories, affect quotas, emotions and feelings. If I imagine that a future SiMA team knows how to integrate the functional ensembles mentioned above in Fig. 6.27 into the three-layered functional picture shown in Fig. 6.14, then all the information is determined that helps a human being make a decision based on their inputs from the outside world, their memories and valuations (quotas of affect, emotions, and feelings). This forces me to conclude that the *free will* can only be an element of imagination, *an illusion*, or as Edwin T. Jaynes sarcastically puts (Jaynes 1993, p. 472) it (freely quoting): "... just as we contemporaries of a late age have no 'free will' unless we believe we have one.".

6.7   Simulation, Experiments, Evaluation and Validation

In computer engineering, the terms simulation, experimentation, evaluation and validation are subject to fixed international definitions and standards, as well as the methods to be used—they are thus legally binding. In other fields, however, and especially in common parlance, the terms are often used in a different sense and the methods are not universally known. For this reason, I would like to address this topic at least briefly. To keep it as simple as possible, I will explain in a few words the scientific contexts as they are used in SiMA. I will avoid profound scientific theories in order to keep it as understandable as possible for readers who have never had to deal seriously with such concepts.

How do you arrive a technical model to begin with? This is well illustrated in the diagram in Fig. 6.28. Experience (empiricism) provides data from the observed behavior of a process. Aristotle already taught how to draw general conclusions from experience and abstraction, i.e. how to derive methods and laws via induction. In contrast to the observed behavior of a natural system, this step leads to a *functional description* of the system in question. The system is thus described in terms of interrelated functions that produce a certain behavior. Deduction, the opposite of induction, helps to draw conclusions from the inferred functional relationships to behavioral patterns.

This loop shows that the *behavior* of a process can only be understood by an abstraction, i.e. the *functional description* always remains an abstraction, no matter

Fig. 6.28 Relationship between functional and behavioral description

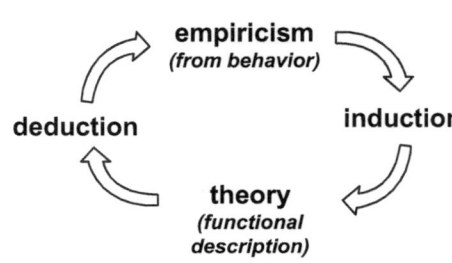

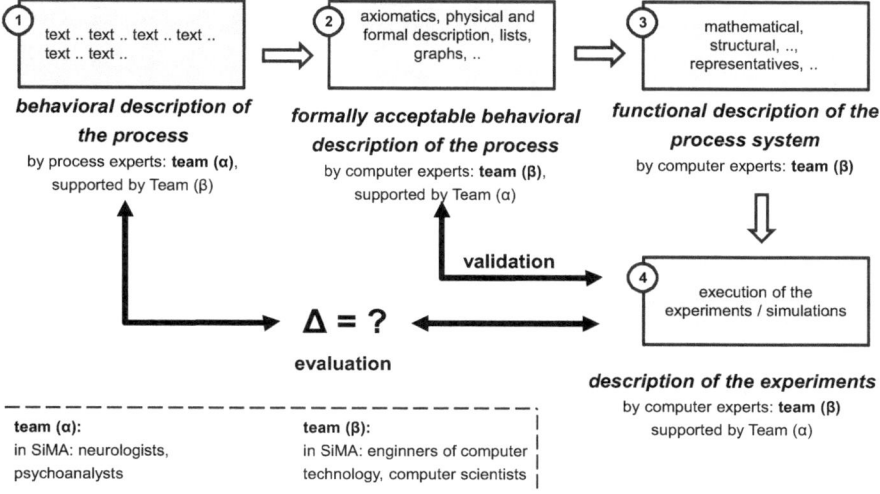

Fig. 6.29 Developmental cycle

how precisely it is formulated. It also shows that, especially in the case of complex behavior, the description will always remain incomplete. Who can fully describe the behavior of a concrete individual, e.g. of me Dietmar Dietrich? And it also shows that the *functional model*, which is the result of the *functional description*, must be checked to see if it is adequately defined for the given object of study.

With this principle of model development in mind, we can proceed directly to the practical implementation in Fig. 6.29. SiMA is based on the theories of the *functional models* of neurology and psychoanalysis (developed by team (α)). Step (1) of Fig. 6.29, the *textual description*, is therefore given by them, but one essential point remains open: In order to be able to test the results later, the experts of step (1) (team (α)) must also describe experiments by which the theoretical model can be tested. Only then does SiMA proceed to step (2). The experts in computer technology and computer science (team (β)) are required to translate the textual descriptions (model and experiments) into an appropriate form with the help of the experts in neurology and psychoanalysis (team (α))—the texts of psychoanalysis today are generally not formulated in a natural scientific, axiomatic form. Step (3) in SiMA has so far been the task of computer technicians and computer scientists, namely the programming of the model or the agents. The agents are the carriers of the simulated Ψ-organ. Step (4) is the execution of simulation experiments. The result of these experiments shows a behavior of the agents, which has to be "examined".

What does "examine" mean in this context? In technology, the corresponding (strictly defined) terms that we use in SiMA are *evaluation* and *validation*.

A first, rough *evaluation* can be done by comparing the result of step (2) with the specifications of step (1). This is only possible if all experts—neurologists, psychoanalysts, computer scientists and computer engineers (team (α) and team (β))—sit down together, reach a consensus and thus jointly accept the result (2). The

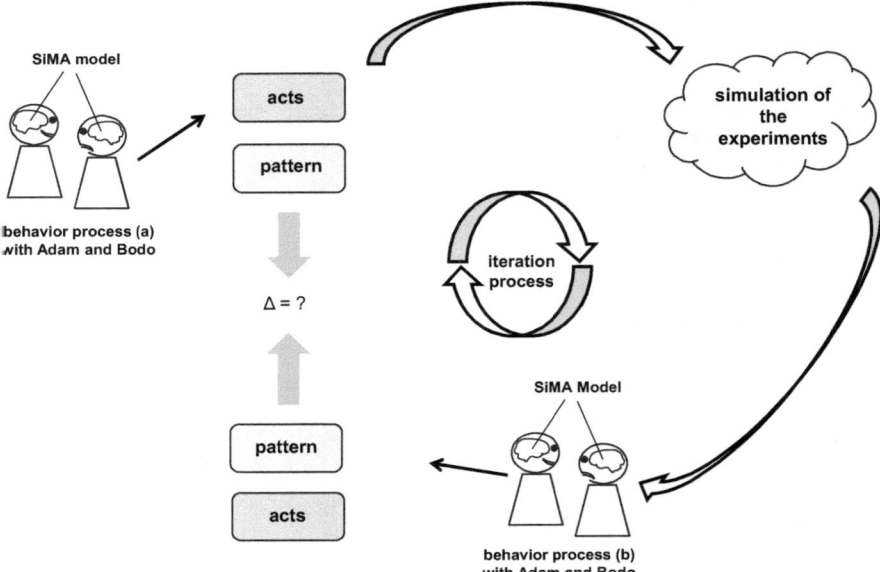

Fig. 6.30 The iteration principle for the optimization of the SiMA three-layer model (Fig. 6.13)

validation is done by the computer engineering experts. This means that the behavior of the experiments in step (4) is *validated* against the specifications of how something should be programmed (2). The last step of this first development process is the real, comprehensive *evaluation*. The neurologists and psychoanalysts must revise the simulation results based on their findings (in Fig. 6.29, step (1)) and check where there are inconsistencies and/or deviations that are no longer acceptable. In other words, are the differences that occur between the process behavior required by the psychoanalysts (1) and the process behavior obtained from the experiments (4), i.e. the aforementioned Δ (delta), either sufficiently good or too large, thus necessitating another round of development iteration?[51]

I would like to illustrate this last train of thought more visually using the illustration in Fig. 6.30. In the first experiments of the SiMA project, the behavior of one or two agents, called Adam and Bodo, is simulated based on the Ψ-organ—the three-level SiMA model of Fig. 6.14. Specific characters are assigned to Adam and Bodo. To keep the initial simulations simple, we imagine the agents at a time

---

[51]Validation, then, is a check that can increasingly be performed using natural scientific methods and tools, while evaluation depends heavily on the opinion of experts in the respective field, which in the case of hermeneutically developed models goes beyond the scope of natural scientific methods. This is the reason why I have included this term in the title of this chapter, since we computer engineers can no longer be responsible for this kind of examination alone.

about 8000 years ago[52] and in a simple, easily defined local environment. The psychoanalysts, who are the experts of the psychic apparatus in SiMA, develop an *act (a)* of how the behavioral process, i.e. a social process between the two agents, could take place.

Here is a simple example. Adam is hungry and sees a piece of meat. He wants to eat it. Then he sees Bodo, who reminds him of his brother, with whom he once haggled fiercely over food. Bad memories are awakened in Adam. Doubts arise as to what is the reasonable thing to do. Should he share the food with Bodo? Should he take it and try to disappear? Bodo can tell from the facial expressions of Adam, whom he does not know, that something is wrong. How should he behave? Adam and Bodo are programmed with various memories of actions that led to success or promised success in similar situations in the past.

The prerequisites for a simulation experiment of the functional model are thus given. The functional structure of the Ψ-organ (the SiMA model) is programmed as far as necessary, the characters of Adam and Bodo are given, the database of memories—i.e. the traces of the life history of the agents—is defined, an environment is determined and thus an initial situation of the presumed behavioral process (of act (a)) devised by the psychoanalysts (top left in Fig. 6.30). On the basis of these specifications, the psychoanalysts can make a rough estimate, based on their experience, of the further course of the act and the patterns to be expected from the functions occurring in the psychic apparatus. The experiment, i.e. the simulation, can be carried out. The functions in the SiMA model (the Ψ-organ of the agents) are activated and generate a behavior (b) (bottom right in Fig. 6.30), which in turn can be described by patterns and the act that takes place. Based on the acts and patterns, the predicted behavioral process (a) can now be compared with the actual behavioral process (b) that resulted from the simulation. The respective acts and the corresponding patterns are the basis for the comparison. It is important to keep this in mind: In the programmed functions of the SiMA models of the agents, monitors are implemented at the crucial points of Fig. 6.14, which continuously record the current states of the validations, such as quotas of affect (pleasure, unpleasure), emotions (joy, anger, fear, ...) and feelings. The expectations of psychoanalysts can thus be checked in detail by measurement and their theories verified. Are their predictions correct, or do certain ideas or even theories need to be corrected?

For acts that are not too simple, interesting differences—measured by Δ—may occur. In such cases, as it is usual in natural scientific experiments, the experimenter must decide whether Δ is still acceptably small or whether the model should be optimized and the previous abstractions improved. It should be clear that the number of iterations required depends on many factors.

Yet there is also the possibility that the experiments were based on incorrect boundary conditions or considerations. This also needs to be checked in case of an unacceptably large Δ. In addition, programming errors may have crept in, but in

---

[52] In SiMA we assume that back then social communication was not as complex as it is today. This greatly simplifies the first experiments in terms of programming effort, which is already enormous.

general these problems—which are annoying but hardly avoidable—are the easiest to detect. Firstly, they often already become apparent during *validation*, and secondly, they are always errors that engineers can track down and isolate.

Three things become clear from such experiments. First, that in principle there can be no mathematical proof of whether the model is correct or incorrect. First, according to Fig. 6.29, such models can only be *validated* and *evaluated*, not proven. In engineering, this is sufficient for such complex processes. I was lucky to be able to work on the Airbus A320 project. There I had to deal with the consequences of a total breakdown of the redundant central computer system[53] of the A320, which inevitably lead to an aircraft crash. I quickly learned that there is no such thing as an absolutely error-free control system (computer). The developer has to live with fact that he is designing an aircraft that will only work with a pre-defined probability. A crash must be considered a residual risk and cannot be ruled out in principle. The residual probability of a crash must be kept as low as possible (Dietrich 1989). All airplanes fly according to this principle, and I think they do it quite well.

Second, the development of the SiMA model, like all models of human organs, is a never-ending story. We are constantly learning. Just think about the simple models of the heart that we had when I was a student and how much knowledge we have gained since then. The amount is enormous. Back then, we were more or less satisfied with being able to record the ECG with three or five electrodes. Today, we can simulate the dynamic electric fields in the heart down to the molecular level—and this in connection with the forces that occur in the heart. In the future, the experiments in Fig. 6.30 will become more complex, more sophisticated, and more broadly based, resulting in an ever-changing picture of the functional model in Fig. 6.14. The many dissertations (see the book's literature) in the SiMA project clearly show this.

Third, such considerations must ultimately make it clear that the SiMA model shown in Fig. 6.14 is one of several possible models. However, as I mentioned earlier, these must not contradict each other. Contrary to the previous approach in SiMA, one could go the way of defining a sub-function called *defense* in the *defense track* and integrate other functions of the id into it, in order to relieve the remaining functions of the defense tracks. This is a thought that I think is worth considering today, as I now consider the defense system to be so complicated and crucial that I would like to subordinate other tasks to these defense mechanisms. However, since this structural definition has no effect on the behavior of the psychic apparatus, but only on the modeling effort, and thus mainly on the software effort, considerations like this are implementation details that I do not want to deal with at this stage of development. At the moment there are much more important tasks than optimizations of this kind.

What is crucial, however, is the consideration of real time, which computer engineers cannot ignore when simulating dynamic processes such as the psychic

---

[53]The central computer system is not to be confused with the autopilot. The autopilot stands hierarchically above the central computer system and sends to it the commands that it has to execute.

apparatus. At SiMA, we have not yet seriously addressed this issue. In the long run, however, it has to be considered. What do I mean by that?

In the natural Ψ-organ, the information received by the sensors makes its way asynchronously through the Ψ-organ until it finally affects the glands and muscles at the output end (see Sect. 5.2). The throughput speed depends on the impulse propagation times in the neurons and the number of successive synapses, as well as certain latencies caused by the psychic apparatus. In the simulation model, which is based on a synchronous rather than an asynchronous switching mechanism—because they are easier to model and because SiMA uses off-the-shelf computers—we developers have to consider a solution to this contradiction. Under no circumstances should a simulation influence the behavior of the simulated processes. But this is a basic knowledge of automation technology, and there are different principles for how to approach this topic, which just need to be taken into account. Simulation has the enormous advantage, as already mentioned, that you can, in principle, extend the time as much as you like. If the developer notices that the processing times of the information influence the behavior of the process, all he has to do is extend the simulation time until the influence largely disappears. However, when the model is actually integrated into a robot, this is no longer possible. The developer then has to deal with the issue of *real-time* in depth. However, as I said—and I studied control engineering—this is above all a task that is solved by investing time.

With regard to simulation and experimentation, there are two other crucial points that are repeatedly raised against SiMA at conferences. I have already touched on both, but with an understanding of how simulation experiments work, one should be able to understand both explanations even better. The first point is that I am a different person now than I was before. I am constantly learning. Every night my dreams revaluate the events of the past day and thus modify my memories. Does that not make it impossible to do repeatable, scientific experiments?

For simulations such as SiMA, this question is easy to answer. All the necessary conditions can be defined, and each function can be observed and recorded in detail using monitors in the Ψ-organ. Experiments can be repeated at will. This means that *well-chosen experiments* help to *evaluate* theories. Concrete individuals like me, for example, cannot be simulated because my history is too vast to be entered into a computer. Nor can I imagine determining the exact structure of the functions of a concrete individual. How should that be possible? The subjectivity of a person remains inaccessible. And the famous human-machine interface, with which "thoughts" could be read out, will remain a huge illusion for the time being, as long as no one is able to develop at least layer L2 (neuronal symbolization) of Fig. 6.6 specifically for a human being. And how this could happen is beyond my comprehension.[54]

---

[54] When individual scientists today claim that they can measure certain activities in the Ψ-organ before they activate a muscle, this means that they can identify a few specific neurons in the Ψ-organ through which the muscle spindles are directly initiated. This does not at all mean that the same method could be used to read thoughts in the future.

Now let us consider the second crucial point: What does "well-chosen experiments" mean in this context? Every experiment requires associations of the psychic apparatus. Associations always depend on the current state, for which all inputs from all sensors (the perception from outside the body, the perception of one's own body, the inner mental state, and above all the state of the current valuation by the quotas of affect, emotions, and feelings) are decisive. This must generally lead to different patterns or inner representatives. If I am sitting happily at a table with friends at home, this generates different patterns in me than if I am separated from my travel group in the jungle and wandering astray at night. This consideration shows that experiments designed to confirm psychoanalytic theories are not easy to choose and can only be developed by a team of experts of the type (α) shown in Fig. 6.29. And something else can be deduced from these considerations: Any theory, no matter how simple it may sound, requires many evaluable experiments before it can be considered confirmed.

6.8 Transition from Model to Physical Realization

Correct! The topic of integrating the SiMA model into a robot, i.e. a physical realization, is not the subject of this book. The main goal was and is to show that neurological models can indeed be linked to psychoanalytical models. Why do I mention the subject of *physical realization* (albeit very briefly)? There are several reasons. One is to show non-engineers that scientific models such as SiMA can of course be translated into physical realizations. It also allows us to see the technical challenges involved in trying to implement this step. I have already touched on this topic several times, but since it is important for AI engineers, I would like to summarize the main points once again.

In the process of technical implementation, three phases must be strictly distinguished: the theoretical model, the simulation model, and the physical implementation. The theoretical model must be considered hypothetical until it has been simulated and physically implemented, since it is developed solely by means of logic and on the basis of natural scientific laws. In the course of such a development, errors of thought, false assumptions, oversights, and other mistakes can always creep in. No one can guard against this. For this reason, when I was a student, such models were first realized on a trial basis. A prototype was developed to show that the model theory was not completely wrong. When computers made it possible in terms of size and cost, engineers switched to simulations, which had many advantages. Costs could be dramatically reduced. Monitors (observation units) could be integrated deep into the model structure to make the smallest technical details visible. Simulations allow us to push physical or chemical experiments to their limits of feasibility or even destruction. And, most importantly for SiMA, simulations can be automated and made automatically observable. Thus, in principle, one can conduct as many experiments as one needs. In practice one can in any case conduct a large amount of experiments and far more variations than if one were using physically and chemically realized objects. And what I particularly like is that we can make the simulation packages available worldwide. Interested parties can often work on them in parallel and contribute without much effort or cost.

Although the scientific work in SiMA has been simulated in many experiments, it has not yet been physically implemented, e.g. integrated into a robot. There are several reasons for this. One reason is that the model has so far been oriented towards translating current psychoanalytic knowledge into natural scientific models, i.e. evaluating the psychoanalytic models on a natural science basis or detecting possible errors in the models. It is necessary to understand the basis, the structure of human intelligence on the basis of natural science. *Then the theories, the models of psychoanalysis, could be evaluated on a natural scientific basis and this knowledge could then be used for the development of AI.*

The idea of integrating the model of the Ψ-organ into a robot still faces a number of challenges. Comprehensive tasks still need to be completed:

– Layer L2 must be developed.
– Layer L1 must be developed, which will have very little to do with that of humans.[55]

Challenges will also include answering specific questions such as:

– Which memories (individual histories) are programmed in which way? The amount of information will be huge in every single case.
– How do you define the relation between drives? Keep in mind that the hormonal system has not been considered yet.

In addition, such a project cannot be simply implemented on the fly, as it requires various specialists, as can be seen from the enumeration above. Projects to realize an artificial Ψ-organ therefore require a very diverse team and several years of development.

And there is one more important point that remains open to me in this context, which is often put forward at conferences as a counter-argument to SiMA. The SiMA model does not address the different brain areas, such as the basal ganglia, mirror neurons, etc. in any way. So, how can we claim in the SiMA project that we are simulating the Ψ-organ? Any of these terms do not refer to the description of the psychic apparatus, but to neurological aspects—i.e. aspects of layer L1—which must also be described with the appropriate neurological methods. For the time being only neurological dummy functions are integrated into the Ψ-organ in SiMA—the L1-functions are only taken into account to the extent that is directly necessary for layer L3, the psychic apparatus.

There is a serious reason why layer L3 was modeled first and not L1 or L2. As explained in Chap. 5, specifying a clear top-down design is a must if you want to

---

[55]This is also one of the reasons why a robot will never feel like a human, which many people refuse to believe because they do not understand the connection between the hardware and the information layers.

build on Mealy's theory. There is only one exception to this rule. If layer L1 (hardware) provides functions whose effects in layer L3 (psychic apparatus) can only be interpreted as complex behavior patterns, one should try to take them into account in the psychic apparatus from the beginning. Mirror neurons are a good example. Before they were discovered, people tried to explain the effect in the psychic apparatus in a different way. Now it is understood, and the effect of mirror neurons can be taken into account directly as corresponding input information of layers L2 and L3 in the model design of the Ψ-organ.

Mirror neurons are thus also a good example of the fact that the top-down design method should not be interpreted to mean that research should not be done in the bottom layer as well. They show that it is best to conduct research on all levels. You just have to understand how it all fits together.

Chapter 7
Conclusion

To help the reader understand the implications of the SiMA project I would like to give a brief overview over the project SiMA.

The Ψ-organ guides us. It develops our knowledge. It develops our insights. What and how we explore, learn, work, think and act depends on it. It is not only the social conditions around us that determine our actions, but also our mental history and our current physiological and mental state. The Ψ-organ is the central unit that we need to understand, for the simple reason that we want to use it to understand everything else. If we do not understand the workings of the Ψ-organ, it is difficult to properly judge other insights. Consider the subject of education. As a child, when I "would not" memorize English vocabulary, I was punished with beatings. And I was certainly not the only one to experience this kind of treatment. Today we know that this method of education cannot create an optimal learning situation. It was the result of the spirit of the times, its zeitgeist, and its logic. But what is the logic of the zeitgeist, the prevailing moral and philosophical ideas? These have often enough led societies into dead ends. That is why *I want to use scientific methods to understand how the Ψ-organ actually works.* Why do we behave in a certain way and not differently? What psychic *functions* control my *behavior*? I must not be fooled by pure logic. Logic can help to point out contradictions. *But only through natural scientific experiments can we work out and evaluate natural laws about our nature.* This is one important insight I have developed in my life as a scientist. The other is crucial to my research in SiMA: the chosen method. Psychoanalysts, neurologists, and AI scientists—in fact, scientists in general—keep to themselves. Real interdisciplinary discussions rarely take place, and when they do, almost everyone concentrates on their own findings, on their own field of knowledge. There is rarely an exchange of experiences where everyone leaves the box of their own field. This has been my experience at many interdisciplinary workshops and conferences,

D. Dietrich, V. Hartmann Cardelle, *Simulating the Mind II*,
https://doi.org/10.1007/978-3-031-69530-8_7

especially those I have organized myself.[1] The experience has almost always been disappointing. This behavior of not wanting to engage with other disciplines can also be seen in another area: Research funding can only be obtained if one is deeply committed to the tenets of one's own discipline, or at least pretends to be. The peer review process leaves little room for anything else. The problem is, however, and we must be clear about this, that it is extremely difficult to do research on the Ψ-organ within the confines of a single discipline, independently of others. Even medicine itself had to accept more and more after the Second World War that methods from mathematics, electrical engineering and computer engineering and computer science can be helpful and often even necessary in order to gain comprehensive knowledge. It is not about messing up each other's works. Complex areas of investigation can only be researched through *interdisciplinary teamwork*—an old wisdom, but one that does not seem to have reached brain research yet. It is no longer possible for one person to cover all the topics in brain research. Too many different areas of expertise are needed. You cannot learn them all in a single lifetime, you would need several lifetimes—unfortunately. Intelligent, committed researchers may be able to acquire the knowledge of several studies, but certainly not the associated skills and the feeling of an expert for the various scientific methods. I had to learn this the hard way at SiMA, even with my best and most highly motivated team members. Everybody needed a lot of time to get a good understanding of the comprehensive topic of *simulation the Ψ-organ*. In the first step one had to apply *the extended theory of Georg H. Mealy*, and after that one needs to correctly use *psychoanalysis and neurology*. And I hope that I have been able to convey the scientific explanation of why this is the case in a reasonably understandable way.

But what are the key issues in these disciplines that are barriers to smooth collaboration?

Psychoanalysts have enormous difficulty accepting a jargon with clearly defined terms, i.e. they have difficulty accepting a clear axiomatics. In clinical work this is simply not possible. From a scientific point of view, however, it is absolutely necessary. Many psychoanalysts also believe that psychoanalysis thrives on contradiction, whereas natural science does not. Thus formulated, this is nonsense. Sigmund Freud never put it that way. It is true that pieces of information often contradict each other, even and especially in the psychic apparatus. *That is the nature of information.* Every automation engineer has to deal with this phenomenon and learn the appropriate principles to deal with this problem. However, this has nothing to do with the fact that the functional structures must be clearly defined and free of contradictions.

Many psychoanalysts fear that the psychic apparatus will be forced into a mathematical framework by the natural scientific way of thinking. I hope that I have been able to dispel this absurd prejudice sufficiently. The structure of the

---

[1]Compare the contributions in the book (Dietrich et al. 2009), which was the result of the conference I organized in Vienna in 2007. The goal of this conference was to bring together scientists from different academic fields.

Ψ-organ model in Fig. 6.14 can be understood without a solid mathematical foundation.

Psychoanalysts have to realize that modern information engineering has passed them by. They work partly with methods of purely textual communication, as Sigmund Freud (mostly) did 100 years ago. They were characterized by complicated textual descriptions and had very few graphics and tables. Descriptions of functions, behaviors, and information, by contrast, must be strictly differentiated, and the interfaces between the functions must be clearly worked out. Is it not much easier to get an overview and understanding of the incredibly complex relationships in a graphic like Fig. 6.14? And even this could be simplified considerably if standardized graphical symbols were introduced for psychoanalysis. We have already tried this to some extent in Fig. 6.14.

As I have seen time and time again at conferences, *neuroscientist* have an enormous fear of people—I repeat myself—messing with their business. They believe that anything that does not come from them can only be a wrong approach. People have often tried to explain this to me. And I am talking about internationally renowned personalities whom I have mentioned several times in the text and whom I hold in high esteem despite these experiences.

Neuroscientists have great difficulty coming to terms with the fact that layer L3 (the psychic apparatus) must be *described* functionally independent of layer L1 (neuronal system). It is true that the sensors record physical quantities, but these only appear as representatives in layer L3. They are no longer physical quantities. If I touch my cold coffee cup, my psychic apparatus does not receive the information that it is 19 °C, but the psychic apparatus receives the representational quantity as information (as a symbol—not as a physical measured quantity): "much too cold" and uses this symbol to associate that this cannot taste good. The transformation from physical to symbolic representation takes place in layer L2. For the structure of layer L3 it is not important how layers L1 and L2 are functionally structured. The psychic apparatus (layer L3) makes demands on layer L2, which it must fulfill. And whether layer L1 is based on the threshold logic or the electronic circuit logic is irrelevant. But this is not true if we think about the sensors (which again are independent of the layer structures). There, it is very much a question of how the information is physically and chemically recorded. This is also one of the reasons why a robot made of sheet metal and silicon *will never feel like a human being*. It will perceive sheet metal and not warm, living skin. It will have different experiences from us humans and therefore create different chains of associations. *Its consciousness will not be humanoid.* A robot will have a machine consciousness.

Natural scientists and *technicians* in particular have the problem of accepting that the majority of psychic functions in primary and secondary processes operate with unconscious information. In abstract and theoretical terms, many can accept this, but when it comes to working out the specific details, I always find that there is enormous resistance. The information theorists of technical engineering must learn to understand that they have by far no knowledge, no methodologically sound psychological experience to understand and properly judge the construct of the psychic apparatus so easily. And they will not have it, not even in the future, because they have never learned let alone studied it in depth. The psychoanalyst is and remains the expert in this field. Physicists, chemists, and technicians must accept this and not constantly question fundamental knowledge or, as often happens, even textbook knowledge.

And that is the problem for us natural scientists (including myself). We too often and too quickly believe that we understand many things because of our training and knowledge. Thus, we computer engineers and computer scientists have to learn to accept and focus on what we can do and what we have always done: *We are experts at translating the models of other sciences into a descriptive language* so that they can be programmed and simulated. It is then the experts of the respective subject-specific models—and not we computer engineers—who can draw their conclusions from these models by means of experiments. In this process, which has been observed since the beginning of computer development, the same thing happens over and over again: The new way of describing information engineering makes it possible to look at models from new perspectives. New, diverse insights can become visible through new, different approaches. That is the nature of this work. However, such findings must always be discussed with the respective experts. This is only possible with them, otherwise it happens much too quickly that wrong conclusions are drawn from these results. We computer engineers then quickly fall into the trap of believing that we can do it ourselves. We all too often imagine that we are smarter than the experts in the other sciences after the first experiments. It is necessary to stay in touch with the *real experts of the models*. That is why I was really pleased when Klaus Doblhammer—and now, with the English translation of the book, Volker Hartmann Cardelle—agreed to constantly check my book for incorrect formulations regarding psychoanalysis. And there were quite a few things I had to correct. These arguments, these discussions between us helped me immensely, and I see the special appeal of the scientific field of computer engineering in these kinds of conversations. It is and always will be incredibly fascinating for me, because I continuously had to understand information systems from other disciplines, even if I could never become an expert in them.

So, if you want to know what the *consequences* of the SiMA project are, it is certainly that research on the Ψ-organ cannot be done by experts from *one* scientific field alone, but that the team *must* include at least psychoanalysts, neurologists, computer engineers and computer scientists. I was able to convince myself of this during the last 24 years in the team, at many conferences, lectures and workshops.

The problem expressed by these considerations meant that my assistants and project engineers, who also wanted to write their dissertation as part of their research, had to overcome at least four major obstacles. First, they had to have spent years studying the theory of computer technology in order to gain an understanding of the extended Mealy theory. Second, they had to understand the distinction between the methods and laws of physics and the methods and laws of information engineering. They needed to understand the distinction between *function* and *behavior*, the connectivity between *signals* and *information* across the layer model, to the point where they had a "gut feeling" for it. Third, they had to learn how to interact on an equal footing with experts from a field that was completely different from their own.

This is something that every computer engineer almost inevitably learns over the years, since computers must now be integrated into all kinds of disciplines. Yet, it has to be learned. Fourth, they must develop a feel for psychoanalysis and neurology. Only when they had overcome these hurdles step by step—and this always took years—were they able to formulate their own scientific task in such a way that they could identify with it. Only then were they able to partially free themselves from the chains of the previous approach to AI in order to sufficiently understand the functioning (not the behavior!) of the human Ψ-organ.

What else do these statements imply? If we manage to bring together the methods and tools of technical information engineering, which have developed enormously through theoretical and practical computer engineering, with the knowledge of psychoanalysis and neurology, we will not only open a wide horizon of thinking to understand the complex interrelationships more easily, but we will also manage to free ourselves from a "mechanistic" way of thinking (Luria 1973, p. 26). *By "mechanistic" I mean the attempt to use physical comparisons for the psychic apparatus and not to apply strictly and consistently the description form of information engineering.* Sigmund Freud also fell into this error all too often due to the zeitgeist of his time, although he himself warned against it. I have addressed this point often enough in the book. It is not only wrong but dangerous. It does not lead to a better understanding, but only to a misguided one. It lures us down false paths and blind alleys. Just think of the concept of psychic energy. These false paths must be avoided at all costs. We must break free from the habit of escaping into physical or chemical descriptions when seeking explanations for the psychic apparatus. This is understandable, because this is the world we internalized in our childhood and youth, especially in school. However, there is no way around it, we have to understand and internalize it: *On the one hand there are physical laws and methods of description, and on the other hand there are the laws and methods of description of information engineering.* The two must not be confused. When this happens, it inevitably leads to contradictions. You end up with statements that can be used to confirm everything and nothing.

This leads me to *philosophy*. I dealt with it intensively in the first semesters of my studies. I found and still find it fascinating. I think it is the same for everyone who engages with it seriously. You can learn an enormous amount about logic, and it is fun, a lot of fun. But as I mentioned above, caution is advised. Philosophers are also experts, experts in their field, but generally not experts and scientists in all other fields at the same time. They are not natural scientific experts in mathematics, psychoanalysis, neurology, computer engineering, or computer science—except for those who have devoted many years of intensive professional study to an additional specialist field. All too often, when they get involved in these fields, they make mistakes and completely lose their way. Yes, I mean it: They lose their way! Just think of Karl R. Popper's discrediting judgment on psychoanalysis (which he later retracted) (Doblhammer 2015). It was not only unscientific, but above all lacking in reflection. Nevertheless, hardly anyone will doubt Karl R. Popper's personality.

Philosophers can certainly make well-founded contributions to projects like SiMA. They have learned and studied to consistently reflect on knowledge, to recognize contradictions, to work out and formulate connections more clearly, unambiguously and literally carve out logic, which we engineers do not really learn in our studies. Unfortunately!

Conversely, when we natural scientists try to tackle philosophical questions, we can quickly become superficial, because we are usually not familiar with the underlying history and thus do not have a complete overview of the relevant contexts. In other words: we usually fail to correctly ask the philosophical questions that we want to deal with and make the task too easy for ourselves. I mentioned the book "Hawking in a nutshell", in which Florian Freistetter explains Stephen Hawking's theory and tries to give the reader an understanding of the philosophical thinking behind it. I found it difficult to do this with the book by Stephen Hawking himself. In contrast, Florian Freistetter's book was the first time I understood convincingly that the question of what happened before the Big Bang is a nonsense question. This is exactly how we have to look at the extended theory of Georg H. Mealy. We technicians seem to be unable to communicate this theory in a clear way, which is why interested people seem to hardly understand it. With this statement I would like to encourage others, especially philosophers, to take up this subject and to communicate it in such a way that as many people as possible have a "heureka" experience, as I had with regard to Stephen Hawkins' theory thanks to Florian Freistetter. Then no one will have to relegate psychoanalysis to the realm of mysticism, as many technicians do.

There is a completely different topic that should not be forgotten under the heading of this chapter *Conclusion* and which has only been hinted at so far: The SiMA project provides methods and tools to bring together the neuronal system and the psychic apparatus in a natural scientific model. This makes it possible to evaluate psychoanalytic, psychological, and sociological theories in a natural scientific way, but also to integrate the structures of SiMA into automated systems (such as robots). This stimulates the imagination of some people to such an extent that they fear that these automated systems could eventually reprogram themselves to oppose Homo sapiens and perhaps even subjugate them as slaves. I can understand such fears. However, remember, fire was not only used to cook food, the ax was not only used to cut down trees, and radioactive radiation was not only used for medicinal purposes. All tools need to be used with care and consideration. With regard to SiMA, however, it is clear that we are still a long way from experimentally evaluate and confirm with the SiMA model the extensive and complex theories of psychoanalysis and sociology that currently exist. The possibility of a machine acquiring the ability to think like a human being in order to reprogram itself to pursue goals other than those we humans set for it is therefore still a long, long way off. Nevertheless, as with the development of the atomic bomb (I am thinking of *Der Fall Oppenheimer* (Jungk 1956)), it is clear that we need to deal with such ethical issues in a timely and intensive manner. I have always addressed this in all my lectures. But I also see a responsibility for philosophers. They are the experts in this field and have the relevant expertise. I also see a responsibility for politicians. As an engineer and scientist in the field of computer technology, I see my task primarily in explaining *where* and *how* automation systems with integrated human-like artificial psychological structures can be used in a meaningful way. And I can formulate such an explanation in a few words: *Wherever humans are.* The work of humans can be made increasingly easier, more pleasant and more economical by an AI that is based on human structures. I often ask the question: Can you think of an area in which people work where machines with intelligence could not help them? I do not know of any such area. Homo sapiens learned to make mechanical tools thousands of years ago. Then they learned to make them mechanically and industrially and to use them on a large scale. Finally, they took the step of making them more and more intelligent. Now they are implanting mental abilities from themselves into them and adapting them to themselves in an ever better way. For me, this is also a contribution to supporting an ever-growing world population to ensure their nutrition and care, including medical care. There should be no question that this has to be done in a socially acceptable way. And people like Trump, Putin, Bolsonaro, Orbán or Erdogan are certainly not competent to come up with the right answers, in my opinion.

This brings me to a very interesting and exciting point: The governments of strongly religious countries such as the USA (under Trump), Poland (under Jaroslaw Kaczynski), Poland, Hungary and Turkey are currently strongly opposing any real aid campaigns for refugees. What do the policies of these governments have to do with the ideas of their respective religions? How can such contradictions be understood? Simple answers like "the bad in a person prevails" will not satisfy any scientist.

I think that once someone has understood SiMA, they have to accept one thing: There is no function in the entire psychic apparatus (layer L3) that allows for *free will*. This confirms the opinion of Arthur Schopenhauer. Free will is and remains a wishful fantasy. From a psychoanalytic point of view:

> People make decisions solely on the basis of their memories, previous attempts at solutions, their assessment of the state of the inner and external world, and the valuation of all this information, based on affects, emotions and feelings.

In the primary and secondary process of Fig. 6.14, one searches in vain for a function of a *free will*.

However, there are strong counterarguments to this point. First, one that we all know: The concept of free will is not axiomatically defined. Philosophy in particular has different points of view, not to mention legal and psychological perspectives. Therefore, I consider it tedious and pointless to argue about free will as long as it is not axiomatically defined.

But let us assume for the sake of argument that we define the term in much the same way. It is often argued that our logic compels us to have free will. Children of different cultures learn their language and cultural logic in their social environment. Why did I learn that $1 + 1 = 2$? I think it is easy to understand: I can still see myself in elementary school putting the red and green sticks next to each other and being complimented when I did something right. And as a child, you want to get (almost) everything right. If I said $1 + 1 = 3$, I was sure to be criticized, laughed at, or at least felt like I was being deprived of love. This created the emotional incentive to quickly learn the logic that $1 + 1$ can only be 2. So, logic is something that a person learns slowly. And what I had to learn about the psychic apparatus is the enormous power of its own defensive function to sometimes bend logic enormously or to subjugate it to unconscious motives. In this consequence, what is important for my decisions is not just my logic, but all the components of the psychic apparatus that help it to make a decision *in any case*, even when the contradictions of the unconscious information at hand are overwhelming. And where does that leave free will? As I said, based on the results of the SiMA model, I agree with Arthur Schopenhauer on this point: *Free will does not exist*—it remains a romantic idea.

I have touched on the subject of axiomatics again, and I think I should address a further contradiction once more. I explained very early on that natural scientific research and especially the world of engineering cannot do without a clear language, i.e. clear axiomatics, which is why, for example, standardization bodies around the world spend many millions of dollars every year on it. No car, no computer, no television would be conceivable without the standardization of language, without axiomatics. On the other hand, I quoted Thomas Bauer (2018), who explains very well that no society would be possible without ambiguity, i.e. ambiguous, unclear or vague formulations. How else could you tell the story of a romantic sunset in the desert? How could a boy tell his parents about the last exciting experience he had with his girlfriend? Does this mean that the contradiction between the natural sciences and the humanities will continue? A solution can also be found with SiMA. On a smaller scale, this shows the increasing convergence of the various trades. Four decades ago, no one would have thought that chimney sweeps would have to talk to electricians, butchers to people in the dairy industry, and farmers to AI technicians. Today, they work together in standards committees and agree on terms to jointly define technical building standards, communication networks as well as safety and security measures. This means that their previously chosen and used terminology and definitions—their axioms—are converging. This is a slow and arduous process. However, it is a process, so it is something dynamic. In this way we have many axiomatically different worlds whose cells are becoming more and more intertwined. Nevertheless, we will not be able to make ambiguity disappear. We need it for our social life. The coexistence of *ambiguity* and *axiomatics* will have to exist side by side for the complex human process. Contradiction must remain, and we must learn to live with it.

The mere demand that the engineering sciences and their information theory have increasingly to engage with the humanities—and vice versa—will trigger a paradigm shift on both sides. Unfortunately, I am quite old and can no longer play an active role in this process of development and debate. No doubt, it will remain a fascinating process.

Appendix by Volker Hartmann Cardelle

Throughout the text I have refrained from making comments with only two exceptions. In order to explain the comments that I feel are necessary, I am instead providing an appendix. It is my wish to offer readers from the field of engineering and the natural sciences some comments that might help them to get a rough idea of how they can better imagine the psychoanalytic side of the model developed in this book, or what kind of questions a psychoanalyst typically deals with.

A.1 Projection

In psychoanalysis there is also another meaning of projection that I consider important to mention (as Dietmar Dietrich writes: psychoanalysis suffers from a lack of axiomatics and this is an instructive example), especially since it is of important for the self, which is an important topic (see Sect. 6.5). According to this other meaning, the defense mechanism of projection is to be understood as follows: Projection means that a quality that belongs to the self or a feeling, emotion or quota of affect of oneself is perceived as belonging to another object or person. For the defense function, this means that through projection, a psychic content (e.g. a quality or an emotion) that belongs to the self is passed on by the defense function as a perception of another object (e.g., another person).

Let us consider a fictitious case. An elderly woman, although past retirement age, works as a psychotherapist and as an examiner of young psychotherapists in clinical training. She has repeatedly experienced situations in which she has not felt up to the task of carrying out her profession, but this feeling has been repressed in those moments because it is a very painful reality to face. Then, when she examines a young candidate, the repressed feeling reappears in the primary process, but this time it is not repressed, it is instead projected. As a result, she perceives the feeling and thought of being unfit for clinical work as a quality of the candidate. As a result, she incorrectly judges the candidate as unfit for this kind of work.

This is an extreme case. Projection can happen to anyone, and almost everyone has probably experienced it without realizing it. However, they tend to occur in situations of lesser importance, and most people are able to recognize that their impression may be wrong when further information is perceived that contradicts their initial impression. For this reason, projections tend to become strong and persistent influences on a person's actions only when the person is either sleepy, intoxicated, or suffering from a severe mental pathology. With age, such pathology may become more likely and more severe.

What is important to understand is that the person whose psychic apparatus conducts a projection, does not notice this. Therefore, from the perspective of personal experience, a projection by your psychic apparatus is not something that you do, but something that happens to you.

A.2 On Hormones of the Sexual Drive

The attentive and critical reader may have noticed the following subtle detail. It is explained in the text the hormonal system is not modeled in the SiMA-model as an additional information *system* to the neuronal *system*. In a later passage, however, the topic of hormones of the sexual drive is mentioned. This could cause irritation, which I would like to resolve.

It was assumed by Sigmund Freud (2018, p. 71) that sexual drives are activated in the nervous system by chemical messengers that originate from other parts of the body (i.e. something for which we have the concept of hormones today). In SiMA, the Ψ-organ was modeled with a dummy function in L1 (F39), which receives input from the body in the form of hormones and produces output in L2, which is finally transformed via the neurosymbolization layer of function F40 into the function F64 of the psychic apparatus). From there it progresses further into the psychic apparatus (see Sect. 6.3.5 and Fig. 6.14).

Since the input into function F39 consists of hormones, it was both necessary and possible to model these specific hormones. This is why the text mentions the topic of hormones of the sexual drive. However, the hormonal *system* is not yet modeled. I hope this clarifies any confusion that may have arisen.

A.3 The Superego Proactive

There was a problem I had with the instance of the superego proactive in Sigmund Freud's structural model that was ironically resolved by the SiMA model. I want to explain this here because I think it is a good example of how SiMA can contribute to psychoanalytic theory.

When I was studying Sigmund Freud, in my understanding his stance on the question of the relationship between the psychic apparatus and the nervous system never felt complete although I was able to name the tenets it consisted of. Since it was only through the Mealy principle that I fully grasped this stance, I rephrase the Mealy principle here to explain the problem I had back then.

The Mealy principle forces any person who considers themselves a natural scientist or materialist to perform some "intellectual acrobatics" (to use Dietmar Dietrich's words in Sect. 4.1) that is not only difficult to perform, but also difficult to accept: One has to accept that there is an entity of scientific description that is not described by physical and chemical laws and methods. This is a hard pill to swallow for anyone who has the fundamental conviction—as I do—that everything in the world is ultimately composed of physically describable matter. This fundamental conviction is not violated, however, because it is not negated that the psychic apparatus ultimately exists in a realized and therefore material form. We are simply unable to understand this physically realized form. And therefore we need the description of the psychic apparatus to be able to understand the Ψ-organ.

Having swallowed this pill one can start to arrange oneself with the model and even embrace it, as one realizes that the model is only driven by inputs from the sense organs and the body via the drives. Thus, the model promises to establish a relationship between any and all psychic contents and processes on the one hand and processes in the body on the other. This is ultimately the goal of a natural science of mental life.

But with this promise in mind, I was somewhat irritated to find that there is an instance that creates activity within the psychic apparatus for action without being stimulated by drive representatives: the superego proactive. Nevertheless, from an empirical point of view (i.e. from the empirical analysis of mental phenomena as conducted by psychoanalysis), it is very reasonable and ultimately necessary to postulate the superego proactive. For me, this had always been a problem with Sigmund Freud's structural model, and I had been eager to find a way to develop a way to link the superego proactive to a drive. To my great astonishment, not only was this not a problem for Dietmar Dietrich as a computer engineer, but he told me that the SiMA experiments had shown that this way of modeling the *superego proactive* seemed to be correct.

This may serve as one example out of many instances in which Dietmar has explained to me that aspects of Sigmund Freud's theory that I had considered weak in terms of their connection to the natural sciences were in fact not a problem.

Ironically, however, I still think that the question of whether the *superego proactive* could be based on a drive is worth investigating. This would require a joint effort by a team formed according to the principles and rules of SiMA and comprising the enumerated fields of knowledge and expertise. I mention this, because I want to show that an objection to details of the model does not necessarily mean that the model, and even less the whole project, has to be rejected. An objection may very well lead to a proposal for a possible adaptation of the model. This means that various aspects of the model with which one feels uncomfortable are either pills to be swallowed or imperfections to be corrected—in either case: they are knowledge to be gained.

But for the time being, the most important task is to work out consciousness. This is a big task that requires the combined efforts of experts in all fields. Until this is achieved, questions of personal interest, such as the superego proactive, will have to wait.

References

(Autor nicht genannt): Das Maximum im Minihirn; Tagesspiegel, Wissen Forschen; Artikel 3 von 5, S. 21; 25.5.2018

Akhter, N.: Visual Tracking of Mechanically Thrown Objects with Planar Surfaces; dissertation at the Institute of Computer Technology, Technische Universität Wien (Austria); 2011

Apprich, C.: Die Maschine auf der Couch. Oder: Was ist schon ‹künstlich› an Künstlicher Intelligenz? Zeitschrift für Medienwissenschaft; Heft 21: Künstliche Intelligenzen, Jg. 11, 2019; Nr. 2, S. 20–28

Auerbach, B.: Ethik, Baruch De Spinoza; publiziert von Jazzbee (über Amazon.de, thalia.de); 12. November 2016

Auerbach, B.: Spinoza; Kindle;15. Sept. 2017

Baars, B.: A Cognitive Theory of Consciousness; Cambridge University Press, 1989

Bammé, A.; Feuerstein, G.; Genth, R.; Holling, E.; Kahle, R.; Kempin, P.: Maschinen-Menschen Mensch-Maschinen, rororo Rowohlt; 1983

Barteit, D. F.: Tracking of Thrown Objects; dissertation at the Institute of Computer Technology, Technische Universität Wien (Austria); 2010

Bartuska, H.; Buchsbaumer, M.; Mehta, M.; Pawlowsky, G. Wiesnagrotzki, S.: Psychotherapeutische Diagnostik: Leitlinien für den neuen Standard; Leitlinien für den neuen Standard; Springer Wien, New York; 2005

Baruch de Spinoza: Theologisch-politische Abhandlung (Tractus theologico-politicus), vollständige deutsche Ausgabe, Übersetzer: Julius Heinrich von Kirchmann, e-artnow, 2014; ISBN 978-80-268-0856-5

Bauer, Th.: Die Vereindeutung der Welt; über den Verlust an Mehrdeutigkeit und Vielfalt; Reclam; 3. Aufl.; 2018

Bear, M. F.; Connors, B. W.; Paradiso, M. A.: Neurowissenschaften: Ein grundlegendes Lehrbuch für Biologie, Medizin und Psychologie Gebundenes Buch – 1. Juli 2016; Springer Spektrum; 3. Aufl.; 2016

Bedini, S. A; Reti (Hrsg.), L.: Leonardo (Hrsg.): Forscher, Künstler, Magier, EMB-Service-Verleger, Bassermann Verlag; 2005

Bethge, P.; Dworschak, M.: Da ist niemand zu Hause; Spiegelgespräch mit David Gelernter; Der Spiegel (8/2011); S. 132

Blum, M.; Williams, R.; Juba, B.; Humphrey, M.: Toward a High-level Definition of Consciousness; Invited Talk to the Annual IEEE Computational Complexity Conference, San Jose CA; 2005

Brainin, E.; Dietrich, D.; Kastner, W.; Palensky, P.; Roesener, C.: Neuro-bionic Architecture of Automation Systems – Obstacles and Challenges; IEEE Africon 2004, Gaborone, Botswana, IEEE Catalog Number: 04CH37590; ISBN 0-7803-8605-1; pp. 1219-1222

Braitenberg, Valentin v.: Vehicles: Experiments in Synthetic Psychology. MIT Press, Cambridge 1984, ISBN 978-0-262-52112-3

Brandstätter, Chr.: Brain-like Location Awareness and Navigation for Psychoanalytically Inspired Autonomous Agents; dissertation at the Institute of Computer Technology, Technische Universität Wien (Austria); 2020

Breazeal, C. L.: Designing Sociable Robots; Massachusetts Institute of Technology, A Bradford Book, The MIT Press Cambridge; 2002

Breuer, J.; Freud, S.: Studien über Hysterie; in: Sigmund Freud Gesamtausgabe (Band 5, S. 131–410); Gießen: Psychosozial-Verlag, 2015

Buddingh, B. C.; Hest J. C. M.: Artificial Cells: Synthetic Compartments with Life-like Functionality and Adaptivity. Accounts of Chemical Research, April, 2017, **50** (4), 769–777

Burghart, C.; Mikut, R.; Asfour, T.; Schmid, A.; Kraft, F.; Schrempf, O.; Holzapfel, H.; Stiefelhagen, R.; Swerdlow, A.; Bretthauer, G.; Dillmann, R.: Kognitive Architekturen für humanoide Roboter: Anforderungen, Überblick und Vergleich; Conference: Proc., 17. Workshop Computational Intelligence, Universitätsverlag Karlsruhe At: Dortmund; 2007

Churchland, P. S.; Sejnowski, T. J.: Grundlagen zur Neuroinformatik und Neurobiologie; Vieweg, Computational Intelligence; 1997

Damasio, A. R.: Descartes' Error, Emorion, Reason and the Human Brain; G. P. Putnam's Son New York, 1994

Damasio, A. R.: The Feeling of What Happens, Body and Emotion in the Making of Consciousness. Harcourt Brace & Company, New York, 1999

Damasio, A. R.: Looking for Spinoza, Joy, Sorrow, and the Feeling Brain; William Heinemann : London; 2003

Damasio, A. R.: Self Comes to Mind. Constructing the Conscious Brain; Pantheon Books, New York; 2010

Deamer, D.: A giant step towards artificial life? Trends in Biotechnology, July, 2005; 23 (7): 336–338

Deutsch, D.: The Psychology of Music; Academic Press, 3rd edition, 2013

Deutsch, T.: Human Bionically Inspired Autonomous Agents; dissertation at the Institute of Computer Technology, Technische Universität Wien (Austria); 2011

Dickerson, R. E.: Chemische Evolution und der Ursprung des Lebens. Spektrum der Wissenschaft. Heft 9, 1979, S. 193

Diener, H. Chr.; Steinmetz, H.; Kastrup, O.: Referenz Neurologie; Thieme; 2019

Dietrich, D., Brandstätter, C., Doblhammer, K., Fittner, M., Fodor, G, Gelbard, F., Huber, M., Jakubec, M., Kollmann, S., Kowarik, D., Schaat, S., Wendt, A., Widholm, R.: Natural Scienific, Psychoanalytical Model of the Psyche for Simulation and Emulation; Scientific Report III; Technische Universität Wien (Austria); 2015

Dietrich, D.: Fehlerredundante Rechner, die Natur als Vorbild; Elektroniker, Nr. 19, Seite 81; 1984

Dietrich, D.: Redundanzprinzipien in Rechnereinheiten, Elektroniker Nr.10/11, Seite 93-99; 1989

Dietrich, D.; Fodor, G.; Pratl, G.; Solms, M.: ENF 2007 - Conference Proceedings; 1st International Engineering & Neuro-Psychoanalysis Forum; Vienna, Austria; IEEE/IEC/INDIN; 2007

Dietrich, D.; Fodor, G.; Zucker, G.; Bruckner, D.: Simulating the Mind – A Technical Neuropsychoanalytical Approach; Springer Wien, New York; 2009

Dietrich, D.; Palensky, P.; Dietrich, D.: Psychoanalyse und Computertechnik eine Win-Win-Situation? psychosozial 35. Jg. (2012) Heft I (Nr. 127); S. 123-135

Dietrich, D.; Penning, K.-F.; Wesseling, H.: Rechneradäquate Minimierung boolescher Funktion mittels des Sharp-Produktes; Elektroniker, Nr. 1/2-1990; S. 61-70

Dietrich, D.; Sauter, Th.: Evolution potentials for fieldbus systems, IEEE International Workshop on Factory Communication Systems WFCS 2000; Instituto Superior de Engenharia do Porto, Portugal; 2000

Dietrich, D.; Schweinzer, H. (Mit-Hsgb.): Feldbustechnik in Forschung, Entwicklung und Anwendung; Springer Wien New York; Tagungsband der Fet'97, 13.10.-14.10.1997, Wien

Dietrich, D.; Zucker, G.; Bruckner, D.; Müller, B.: Neue Wege der kognitiven Wissenschaft im Bereich der Automation; Elektrotechnik & Informationstechnik (2008), e&i; 125/5: 180–189. https://doi.org/10.1007/s00502-008-0536-x

Dietrich, D.; Jakubec, M.; Schaat, S.; Doblhammer, K.; Fodor, G.; Brandstaetter, C.: The Fourth Outrage of Man (Is the Turing-Test Still up to Date?), Journal of Computers vol. 12, no. 2, pp. 116-126; 2017a

Dietrich, D.; Zucker, G.; Doblhammer, K.: On the way to bridging the gap between the mental apparatus and the neurobiological layer, Neuropsychoanalysis, 1972, 19:2, 159-173, https://doi.org/10.1080/15294145.2017.1383180; 2017b

Doblhammer, K.: Das Selbst eines Roboters; Jubiläumsstiftung der Stadt Wien für die österreichische Akademie der Wissenschaften; J – 3/12 (Juni 2013); Projekt-Nr. 150115

Doblhammer, K.; Jakubec, M.; Schaat, S.; Dietrich, D.; Widholm, R.; Fodor, G.: Popper Revisited - Epistemological Considerations on Psychoanalysis Regarding a Technical Model of the Psyche; Publikationsdatenbank der Technischen Universität Wien; 2015

Dokaupil, E.: Wenn Technik wie der Mensch denkt; Das Innovationsmagazin inhi!tech von Siemens Österreich; Dezember 2006a, 4/06; S. 52

Dokaupil, E.: Das Gehirn verstehen – Mark Solms; Das Innovationsmagazin inhi!tech von Siemens Österreich; Dezember 2006b, 4/06; S. 54

Dönz, B.: External Semantic Annotation of Web Databases; dissertation at the Institute of Computer Technology, Technische Universität Wien (Austria); 2015

Dornes, M.: Der kompetente Säugling - Die präverbale Entwicklung des Menschen; Geist und Psyche Fischer; Fischer Taschenbuch Verlag; 2001

Eccles, J. C.: Das Gehirn des Menschen; R. Piper u. Co. Verlag, München; 1975.

Eccles, J. C.: How the Self Controls ist Brain, Springer Verlag Berlin, Heidelberg 1994

Erikson, E.: Der vollständige Lebenszyklus; suhrkamp taschenbuch wissenschaft; 10. Auflage; 2018

Erismann, Th. H.: Zwischen Technik und Psychologie, Grundprobleme der Kybernetik; Springer-Verlag, Berlin, Heidelberg, New York; 1968

Fittner, M.: Simulation von menschlichem Lernen in einem Funktionsmodell der menschlichen Psyche; dissertation, Vienna University of Technology (Austria), Institute of Computer Technology; voraussichtliche Publikation 2021

Foerster, H. v.: Wissen und Gewissen; Suhrkamp taschenbuch wissenschaft 876; 1993

Forth, E.; Schewitzer, E.: Bionik; Meyers Taschenlexikon; VEB Bibliografisches Institut, Leipzig; 1976

Frank, H.: Kybernetik – Brücke zwischen den Wissenschaften; Umschau Verlag Frankfurt am Main, 7. Aufl.; 1970

Frank, H.; Barteit, D. F.; Kupzog, F.: Throwing or shooting - a new technology for logistic chains within production systems; in: IEEE Int. Conf. on Technologies for Practical Robot Applications (TePRA 2008); 2008, p. 62–67

Freistetter, F.: Stephan Hawking: His Science in a Nutshell, Prometheus Books, Illustrated Edition, 2019

Freud, S.: Entwurf einer Psychologie (1895), in: Gesammelte Werke (Nachtragsband, S. 386-477); Frankfurt am Main: Fischer Taschenbuch Verlag; 1999

Freud, S.: Zur Auffassung der Aphasien, ein kritisches Vorwort von W. Leuschner; Studie; Hrsg.: Fischer Taschenbuch Verlag; 2. Aufl.; 2001

Freud, S.: Die Abwehr-Neuro-psychosen; in: Sigmund Freud Gesamtausgabe (Band 4, S. 381-395); Gießen: Psychosozial-Verlag, 2015

Freud, S.: Die Traumdeutung; in: Sigmund Freud Gesamtausgabe (Band 7); Gießen: Psychosozial-Verlag; 2017

Freud, S.: Drei Abhandlungen zur Sexualtheorie; in: Sigmund Freud Gesamtausgabe (Band 10, S. 13-95); Gießen: Psychosozial-Verlag; 2018a

Freud, S.: Formulierungen über die zwei Prinzipien des psychischen Geschehens; in: Sigmund Freud Gesamtausgabe (Band 12, S. 173-182); Gießen: Psychosozial-Verlag; 2018b

Freud, S.: Triebe und Triebschicksale; in: Sigmund Freud Gesamtausgabe (Band 14, S. 231-251); Gießen: Psychosozial-Verlag; 2020a

Freud, S.: Das Unbewußte; in: Sigmund Freud Gesamtausgabe (Band 14, S. 267-301); Gießen: Psychosozial-Verlag; 2020b

Freud, S.: A pszichoanalízis egy nehézségéről [Eine Schwierigkeit der Psychoanalyse]; in: Sigmund Freud Gesamtausgabe (Band 16, S. 15-31); Gießen: Psychosozial-Verlag; 2020c

Freud, S.: Jenseits des Lustprinzips; in: Sigmund Freud Gesamtausgabe (Band 16, S. 395-447); Gießen: Psychosozial-Verlag; 2020d

Freud, S.: Psychoanalyse; in: Sigmund Freud Gesamtausgabe (Band 17, S. 201-222); Gießen: Psychosozial-Verlag; 2020e

Freud, S.: Das Ich und das Es; in: Sigmund Freud Gesamtausgabe (Band 17, S. 269-311); Gießen: Psychosozial-Verlag, 2020f

Freud, S.: Neue Folge der Vorlesungen zur Einführung in die Psychoanalyse; in: Sigmund Freud Gesamtausgabe (Band 19, S. 279-438); Gießen: Psychosozial-Verlag; 2021a

Freud, S.: Abriss der Psychoanalyse; in: Sigmund Freud Gesamtausgabe (Band 20, S. 367-484); Gießen: Psychosozial-Verlag; 2021b

Friedman, L.: Letters to the Editor. Int. J. Psycho-Anal., 56:123-127; 1975

Gabbard, Glen O.: Litowitz, Bonnie E.; Williams, Paul: Textbook of Psychoanalysis, Second Edition, American Psychiatric Publishing; 2012

Gelbard, F.: Psychoanalytic Defense Mechanisms in Artificial Intelligence Exemplified Implementation in Complex Technical Systems; dissertation, Vienna University of Technology (Austria), Institute of Computer Technology; 2015

Gill, M. M.: Metapsychology is not Psychology, in Gill/Holzmann: Psychology is Metapsychology: Essays in Memory of George Klein, New York; 1976

Goldscheider, P.; Zemanek, H.: Computer, Werkzeug der Information; Springer-Verlag, Berlin Heidelberg New York; 1971

Green, A.: Analytiker, Symbolisierung und Abwesenheit im Rahmen der psychoanalytischen Situation; Psyche 29, S. 503-541; 1975

Harari, Y. N.: Homo Deus. A Brief History of Tomorrow; Publisher: Harvill Secker; 2015

Harlfinger, J.: Jeder braucht eine besondere Therapie - Wer sind heute die wahre Erben Sigmund Freuds; Der Standard, MEDSTANDARD; 15.10.2007, S. 20

Hartmann Cardelle, V.: Metapsychological consequences of the conscious brainstem: A critique of the conscious id; Neuropsychoanalysis; 2019

Hartmann Cardelle, V.; Dietrich, D.: Understanding metapsychology with the computer paradigm, Metapsychologica – Rivista di psicanalisi freudiana, no.1 (2022); pp. 137-164

Hartmann, H. (1950): Comments on the Psychoanalytic Theory of the Ego. Dt: Bemerkungen zur psychoanalytischen Theorie des Ichs; Psyche; 1964, 18, S. 330-354

Hasler, F.: Neuromythologie; Eine Streitschrift gegen die Deutungsmacht der Hirnforschung; [transcript] XTEXTE; 2012

Hassenstein, B.: Biologische Kybernetik; Quelle & Meyer Heidelberg; 1973

Hawkins, J.: On Intelligence. Henry Holt and Company. New York; 2004

Hinterhuber, H.: Die Seele. Springer Verlag, Wien – New York; 2001

Holt, R.R.: The Crisis on Metapsychology. Paper read at the American Academy of Psychoanalysis; quoted after: (1997). References for Dr. Rangell's Presentations. J. Clin. Psychoanal., 6(4): 588; 1973

Holt, R.R.: The Current Status of Psychoanalytic Theory. Psychoanal. Psychology, 2:289-315; 1985

Huber, M.: Die Bedeutung von Emotion für Entscheidung und Bewusstsein; Die neurowissenschaftliche Herausforderung der Pädagogik am Beispiel von Damasios Theorie der Emotion; Verlag Königshausen & Neumann; 2012

Humanoiden nehmen erst 2050 Vernunft an; Autonome Servicesysteme auf zwei Beinen besitzen Potential, Computer Zeitung, Nr.32-33; 11.8.2003

Irrgang, B.: Posthumanes Menschsein? – Künstliche Intelligenz, Cyberspace, Roboter, Cyborgs und Designer-Menschen – Anthropologie des künstlichen Menschen im 21. Jahrhundert; Franz Steiner Verlag; 2005

Jacobson, E.: Das Selbst und die Welt der Objekte; suhrkamp taschenbuch wissenschaft; 5. Aufl.; 1998

Jaynes, J.: Der Ursprung des Bewusstseins; Rowohlt Taschenbuch Verlag GmbH; 1993

Jung, R.; Kornhuber, H.: Neurophysiologie und Psychophysik des Visuellen Systems; Symposion Freiburg/BR., 28.8.-3.9.1960; Springer Verlag; 2013

Jungk, R.: Heller als tausend Sonnen. Das Schicksal der Atomforscher, rowolt; 1956

Jürgensohn, Th.: Psycho-Ingenieurwesen - Hinweis für interdisziplinär arbeitende Ingenieure; atp 48, Heft 7; 2006; S. 48 - 58

Kabitzsch, K.; Dietrich, D.; Pratl, G.: LONWORKS – Gewerkeübergreifende Systeme; VDE Verlag GmbH Berlin, Offenbach; 2002

Kandel. E. R.: In Search of Memory: The Emergence of a New Science of Mind; W. W. Norton & Co.; 2006

Kandel. E. R.: Psychiatry, Psychoanalysis, and the New Biology of Mind; American Psychiatric Publishing; 2008

Kaplan-Solms, K.; Solms, M.: Clinical Studies in Neuro-Psychoanalysis: Introduction to a Depth Neuropsychology; Other Press; 2nd edition; 2001

Kernberg, O.: Object Relations Theory and Clinical Psychoanalysis; Jason Aronson Inc. Publishers; 1995

Klaus, G.: Wörterbuch der Kybernetik; Fischer Handbücher, 1969

Klein, D.: Vom Tanzen der Seele; Zeit Magazin; Nr. 50, S. 22-28; 10. Dezember 2015

Küpfmüller, K.; Jenik, F.: Über die Nachrichtenverarbeitung in der Nervenzelle; Kybernetik, Band I, Heft 1; Januar 1961

Lang, R.: A Decision Unit for Autonomous Agents Based on Modern Psychoanalysis; dissertation at the Institute of Computer Technology, Technische Universität Wien (Austria); 2010

Lange, D.: Aufbau zweier Neuronenmodelle mit modernen Bauelementen nach Küpfmüller-Jenik zur Nachbildung gegenseitiger Erregungs- und Hemmvorgänge bei neuronaler Verkopplung über exzitatorische beziehungsweise inhibitorische Synapsen; Diplomarbeit an der TU Karlsruhe, Fakultät für Elektrotechnik, Institut für Kybernetik; 1976

Leuzinger-Bohleber, M.: "Embodied" Memories – Vergangene Traumatisierungen in gegenwärtiger Inszenierung; psychosozial; ISSN 0171.3434.31. Jahrgang Nr. 111; 2008, Heft 1, S. 27

List, E.: Psychoanalyse- Geschichte Theorien Anwendungen; facultas wuv, UTB; 2009

Loy, D.; Dietrich, D.; Schweinzer, H.-J.; Open Control Networks LonWorks/EIA 709 Technology; Kluwer Academic Publishers; 2001

Lunze, J.: Künstliche Intelligenz für Ingenieure, Technische Anwendungen; R. Oldenbourg Verlag München Wien; 1995

Luria, A. R.: The Working Brain – An Introduction to Neuropsychology; BASIC BOOKS, A Member of The Perseus Books Group; 1973

Luria, A. R.: Das Gehirn in Aktion. Einführung in die Neuropsychologie. rororo science, 6. Aufl.; Oktober 2001

Martin, H.: Wenn es Krieg gibt, gehen wir in die Wüste. Verlag der Namibia Wissenschaftlichen Gesellschaft, Windhoek, Namibia; 2009

Mauthner, F.: Spinoza, Ein Umriss seines Lebens und Wirkens (vollständige Biografie); e-artnow; 2015; ISBN 978-80-268-2809-9

Mealy, G. H.: A Method for Synthesizing Sequential Circuits, Bell System Tech. J. 34, pp. 1045–1079; September 1955

Meyer-Waarden, K.; Einführung in die biologische und medizinische Meßtechnik, F. K. Schattauer Verlag Stuttgart – New York; 1975

Meyer-Waarden, K.; Lange, D.; May, H. U.: Halbautomatische elektronische Steuerung eines Mikroelektroden-Vortriebes; Meyer-Waarden, K.; Lange, D.; May, H. U.; Biomedizinische Technik, Band 20, H. 2/1975; S. 71-76

Mironov, K.: Predicting the Trajectory of the Flying Object with the Use of k-Nearest Neighbors; dissertation, Technische Universität Wien (Austria), Institute of Computer Technology; 2016

Muchitsch, C.: Human-like Perception for Psychoanalytically Inspired Reasoning Units; dissertation at the Institute of Computer Technology, Technische Universität Wien (Austria); 2013

Palensky, B.: Introducing Neuro-Psychoanalysis towards the Design of Cognitive and Affective Automation Systems; dissertation at the Institute of Computer Technology, Technische Universität Wien (Austria); 2008

Panksepp, J.: Affective Neuroscience. The Foundations of Human and Animal Emotions. Oxford University Press.1998

Patestas, M. A.; Gartner, L.: A Textbook of Neuroanatomy; Hoboken, New Jersey: John Wiley & Sons Inc.; 2016

Peterfreund, E.: On Information and Systems Models for Psychoanalysis. Int. Review of Psycho-Analysis, 7:327-345; 1980

Peterfreund, E.: On information-processing models for mental phenomena - a reply to Lawrence Friedman's critical review of information, systems, and psychoanalysis by Emanuel Peterfreund; The International Journal of Psychoanalysis; 1973, 54, 351–357.

Peterfreund, E.; Schwartz, J.: Information, systems, and psychoanalysis. An evolutionary biological approach to psychoanalytic theory. New York: International University Press; 1971

Pongratz, M.: Bio-inspired Transport by Throwing System; dissertation, Technische Universität Wien (Austria), Institute of Computer Technology; 2016

Popper, K. R.; Eccles, J. C.: The Self and Its Brain - An Argument for Interactionism; Springer Verlag Heidelberg, Berlin, London, New York; 1977

Powell, K.: How biologists are creating life-like cells from scratch; springer nature; Nov. 2018

Pratl, G.: Processing and Symbolization of Ambient Sensor Data; dissertation, Technische Universität Wien (Austria), Institute of Computer Technology; 2006

Reppen, J.: Symposium: Emanuel Peterfreund on Information and Systems Theory. Psychoanal. Rev., 68:159-161; 1981

Ridley, M.: The Rational Optimist. How Prosperity Evolves; Fourth Estate, London; 2010

Rubner, J.: Intelligenz in Robotergehirnen. Pictures of the Future. Die Zeitschrift für Forschung und Innovation. Siemens; Herbst 2002. Seite 66-68

Russell B., Whitehead A.N.: Principia Mathematica, 3 volumes, Merchant Books, 2009

Sacks, O.: The Man Who Mistook His Wife for a Hat; Summit Books/Simon & Schuster, Inc., New York; 1985

Sandler J.S.; Sandler A.-M.: Internal Object Relations; Routledge, London; 1998

Schaat, S.: Simulation of Foundational Human Information-Processing in Social Context; dissertation, Technische Universität Wien (Austria), Institute of Computer Technology; 2016

Schank, R. C.; Childers, P. G.: Die Zukunft der künstlichen Intelligenz – Chancen und Risiken; Du Mont; 1986

Schmitz, D.; Schuchmann, S..; Fisahn, A.; Draguhn, A.; Buhl, E. H.; Petrasch-Parwez, E.; Dermietzel, R.; Heinemann, U.; Traub, R. D.: Axo-axonal coupling. a novel mechanism for ultrafast neuronal communication; National Library of Medicine; National Center for Biotechnology Information; Neuron; 2001 Sep 13;31(5):831–40. https://doi.org/10.1016/s0896-6273(01)00410-x

Schülein, J. A.; Die Logik der Psychoanalyse. Psychosozial-Verlag, Gießen; 1999

Shannon, C. E.: A Mathematical Theory of Communication; in: Bell System Technical Journal. Short Hills N.J. 27.1948, (Juli, Oktober), S. 379–423, 623–656; ISSN 0005-8580

Singer, W.: Vom Gehirn zum Bewusstsein; Suhrkamp; 2006

Sloman, A.: Machines in the Ghost, in: Dietrich, D.; Fodor, G.; Zucker, G.; Bruckner, D.: Simulating the Mind – A Technical Neuropsychoanalytical Approach; Springer Wien, New York; 2009; p. 124

Solms, M. (2013). The Conscious Id. Neuropsychoanalysis, 15 (1), pp. 5–19

Solms, M., Turnbull, O.: The Brain and The Inner World, An Introduction to the neuroscience of Subjective experience; Other Press New York; 2002

Solms, M; Friston, K; (2018) How and why consciousness arises: Some considerations from physics and physiology. Journal of Consciousness Studies, 25 (5-6) pp. 202-238

Solso, R. L.: Kognitive Psychologie; Springer, 2005

Steinbuch, K.; Rupprecht, W.: Nachrichtentechnik, Eine einführende Darstellung; Springer-Verlag, Berlin/Heidelberg/New York; 1967

Tan, D. S.; Nijholt, A.: Brain-Computer Interfaces, Applying our Minds to Human-Computer Interaction; Springer; 2010

Turkle, Sh.: Artificial Intelligence and Psychoanalysis: A New Alliance; in the Artificial Intelligence Debate; False Starts, Real Foundations; Edited by Stephen R. Graubard; MIT Press, 1988; S. 242-268

Velik, R.: A Bionic Model for Human-like Machine Perception; dissertation, Technische Universität Wien (Austria), Institute of Computer Technology; 2008

Walke, B.: Datenkommunikation I; Teil 1: Verteilte Systeme, ISO/OSI-Architekturmodell und Bitübertragungsschicht; Hüthig Buch Verlag Heidelberg, 1987

Weizenbaum, J.: Computer Power and Human Reason. From Judgement to Calculation; 1976

Wendt, A.: Experience-Based Decision-Making in a Cognitive Architecture; dissertation at the Institute of Computer Technology, Technische Universität Wien (Austria), 2016

Wendt, S.: Entwurf komplexer Schaltwerke; Springer Verlag; 2013

Wiener, N.: Ich und die Kybernetik; Wilhelm Goldmann Verlag München; 1970

Wieser, B. U.: Chinas barbarische Jugend; Neue Züricher Zeitung, 4.8.2016

Wiest, G.: Hierarchien in Gehirn, Geist und Verhalten; Ein Prinzip neuraler und mentaler Funktion; Springer Wien New York; 2009

Wilhelm, D.: Von der Ordnung im Gedächtnis; Neunkirchener Druckerei und Verlag; 1982

Yalom, I. D.: The Spinoza Problem; Basic Books, New York, 2012

Yovell, Y.: Liebe und andere Krankheiten; btb; 2008

Yovell, Y.: Return of the Zombie – Neuropsychoanalysis, Consciousness, and the Engineering of Psychic Functions, in: Dietrich, D.; Fodor, G.; Zucker, G.; Bruckner, D.: Simulating the Mind – A Technical Neuropsychoanalytical Approach; Springer Wien, New York; 2009, p. 251-258

Zeilinger, H.: Bionically Inspired Information Representation for Embodied Software Agents; dissertation at the Institute of Computer Technology, Technische Universität Wien (Austria), 2010

Zemanek, H.: Claude E. Shannon; it+ti, 4/2001; S. 1-7

Zimmermann, H.: OSI Reference Model – The ISO Model of Architecture for Open Systems Interconnection; IEEE Transaction on Communications, VOL. Com-28; NO.4, April, 1980

Internet Addresses

Blue Brain Project, 2005. https://www.epfl.ch/research/domains/bluebrain/

Bose, 2010. https://slideplayer.com/slide/4741531/ Introduction to the mathematical modeling of neuronal networks; Jawaharlal Nehru University & New Jersey Institute of Technology; IISER, Pune February, 2010

Digressions&Impressions. https://digressionsnimpressions.typepad.com/ digressionsimpressions/2014/07/was-newton-familiar-with-spinoza-1.html

Forschung & Wissen, 2014: https://www.forschung-undwissen.de/nachrichten/biologie/biologen-erschaffen-kuenstliches-leben-13371939

Frontiers, 2018. Frontiers | Human Consciousness: Where Is It From and What Is It for (frontiersin.org)

Herbert A. Simon, 1965: (Stanford) http://www.zeit.de/digital/internet/2016-08/kuenstliche-intelligenz-geschichte-neuronale-netze-deep-learning/seite-2

Honey, 2016. https://www.zeit.de/digital/internet/2016-08/kuenstliche-intelligenz-geschichte-neuronale-netze-deep-learning

Hyperscience, 2023. Hyperscience Wins 2023 Artificial Intelligence Breakthrough Award: https://www.hyperscience.com/newsroom/hyperscience-wins-2023-artificial-intelligence-breakthrough-award/

https://npsa-association.org/who-we-are/

Insects Conscious, 2022. https://www.noemamag.com/the-surprisingly-sophisticated-mind-of-an-insect/

Kriesel, 2005. http://www.dkriesel.com/en/science/neural_networks

Nature Reviews Neuroscience. The brainstem control of saccadic eye movements; https://www.nature.com/articles/nrn986

NIH, 2001. NIH.gov (National Library of Medicine): https://www.ncbi.nlm.nih.gov/books/NBK10885/

Online Lexikon. Online Lexikon für Psychologie und Pädagogik (Stangl, 2020); https://lexikon.stangl.eu/1160/assoziation/

Oxford Handbooks online. http://www.oxfordhandbooks.com/view/10.1093/oxfordhb/9780199930418.001.0001/oxfordhb-9780199930418-e-27

Palm history. https://history-computer.com/palm-history/

Reutetzki, P.: Der Philosoph, dem ich zumeist vertraue – Über Goethes Verhältnis zu Spinoza unter besonderer Berücksichtigung des Faust; Vortrag, Goethe Gesellschaft Erfurt e.V., 2019: https://www.goethe-gesellschaft-erfurt.de/der-philosoph-dem-ich-zumeist-vertraue-ueber-goethes-verhaeltnis-zu-spinoza-unter-besonderer-beruecksichtigung-des-faust/

Science, 2010. Elizabeth Pennisi. https://doi.org/10.1126/science.328.5981.958. Science VOL 328 (5981), 958-959. 21 MAY 2010. http://science.sciencemag.org/content/328/5981/958

Seismart, 2019. http://seismart.de/mehr/tierwelt/haben-ameisen-ein-gehirn-wie-intelligent-sind-ameisen/

slogans, 2018. https://www.slogans.de/slogans.php?BSelect%5B%5D=621

Spektrum, 2018. https://www.spektrum.de/magazin/haben-insekten-bewusstsein/1557082

Spiegel online, 2008. http://www.spiegel.de/wissenschaft/natur/insekten-gehirne-schaben-haben-ein-bewusstsein-a-589476.html

Universitätsklinikum Freiburg, 2019. dasgehirn.info : https://www.dasgehirn.info/wahrnehmen/sehen/hauchduenner-hochleistungsrechner

Watanabe et al., 2014. https://www.uni-heidelberg.de/presse/news2014/pm20141222_auf-der-suche-nach-dem-ursprung-unseres-gehirns.html

Index

A

Abstraction, 56
Abstraction level, 145
Abstraction model, 125
Act, 173, 175, 184, 195, 205
Acting, 176
Actions, trial, 176
Actuator, 6, 29, 72, 95, 139
Address driver, 193
Affect, 19
Agent, 133, 203, 204
Aggression, 73
Aggressiveness, 157, 171
Air traffic, 3
Algorithm, 4
Ambiguity, 84, 218
Analog, 3
Anger, 152, 159, 171
Animal, 135
Ant, 30, 74
Anxiety, 152
Aphasia, 92, 191
Aphasie, 191
App, 5
Apparatus
    mental, 137
    psychic, 137
Artificial intelligence (AI), xxiii, 1, 4, 71, 192
Artificial Recognition System (ARS), xxiii
Associations, 153, 198
    chains of, 151
    corresponding, 48
Asterisk, 165
Asynchronous, 118, 132, 207

Austrian Standards International (ASI), 20
Automation, 1, 3, 92
Awareness, 3
Axiom, 18, 22
Axiomatics, 7–23, 218
Axiomatic terminology, 20

B

Ball, catch a, 49
Basal ganglia, 209
Basic emotion, 59
Bauer, Th., 84
Bee, 30, 74
Behavior, 7, 71, 72, 133, 148, 152, 214
    dynamic, 74
    rules of, 66
Behavioral model, 128, 149
Behavioral phenomena, 16
Behavioral scientist, 159
Behavior based artificial
    intelligence, 13
Benchmark, 164
Biological blow, 17
Biology, 17
Blow, narcissistic, 17
Blows of mankind, xvi
Blue Brain, 7
Body, xi, 10, 152, 195, 198
Bottom-up, 45
Brain, 27
Brainstorming, 62, 153
Bridge, 110
Bridging the gap, 85